Thank You, President Corona!

ALPHA VOLUME
(Volume I)

How COVID-19 Improved the World

By Alex Joonto

Proof reading by Ivana Maglica

For contacts and collaborations:
email alexjoonto@proton.me
Instagram alexjoonto

https://www.thankyoupresidentcorona.com

Acknowledgments

Beside thanking COVID-19, there is a long list of people I am grateful for, without which this book would have never come to life, or would have never been so interesting.

I can't thank them one by one, or this opening would take the whole book. They know who they are.

However, I must mention a few folks that were key in writing down these pages.

I want to thank my dear friend Ivana Maglica, who enthusiastically offered to proofread my work. She was instrumental in turning a shapeless and meaningless mountain of pages, into a book with a structure and a message. I think Ivana still feels sorry for her direct and honest feedback, but in reality, that's why I love her!

Thanks to Ivana, my buddy Mickey Bortel could enjoy a fresh manuscript as a test reader, defining it "fun to read and well researched!"

Special thanks go to the very people I had the privilege to spend the lockdown period with, my flatmates Tunde and Cobra, our landlords/friends James and Karolina, plus their fantastic child Leo, and their fluffy animals: Luna the Siberian husky, and Frody the silver Bengal! I couldn't ask for better companions!

You may call me a weirdo, but I want to also thank all of those folks who hurt me, or to use a more poetic expression, those who broke my heart. Finally the light can get in!

This volume is dedicated to my cat Alfa, who passed away on 2nd September 2022.

Preface

This book is my Meraki. This Greek word describes some work, usually artistic, where you put in all your energy and passion. You do this work with love, and as a result, it gives your life a purpose.

But, you may wonder by now, what the hell is this book about? Why the hell should you yell, "Thank You, President Corona!"???

When the WHO declared the state of pandemic, the world panicked. However, my reaction was much weirder: I felt the urge to do something with my life.

An apocalypse was unraveling. "Hard times make strong men..." I thought. I wanted to survive this pandemic and get out of it as a strong man.

Paranoia spread faster than the virus itself, crumbling our sense of security. The empty streets were the cover of all the plans gone South. This sudden freezing in our lives marked the end of the world as we knew it. But when one world ends, another begins...

The first thing I had forbidden myself to do was to complain. I didn't let myself fall into despair. There was nothing I could complain about. I had kept my job when many others didn't. I was in Malta, an island relatively safe from the infected European continent. Mostly, I wasn't in the risk group. I was just 31, leading a pretty healthy life, though maybe not as healthy as I thought. I've never smoked, never used drugs, and have always been drinking responsibly. What I was missing was sport. I had been going to the gym for a while, but I had taken it so easy... It was only when I began to follow workouts on YouTube during the lockdown, that I started to challenging my body and actually transforming it!

This was the start of a self-development process that will inspire the concept of thanking the virus for all the positive changes I have observed in the last two years. This pathogen has been devastating to many lives, but on the other hand, it has forced us to take measures never seen before and forced us to innovate where we thought it was impossible to innovate, not because of the actual limitations but because of the classic fear of change.

Most people are scared of change, but a virus can be scarier than change! A leader can either lead by example or by fear. The coronavirus led us by fear. It has led us to change our habits, our economy, our ways of working. It has brought us closer just when the world was being ripped apart by a more and more polarized society. This made me feel as if the coronavirus was a true, though invisible, leader, the first president of the world! President Corona!

2020 was the most productive year of my life! Who did I have to thank for the most productive year of my life? President Corona!

Who did I have to thank for working from home? President Corona!

Who pushed me to work out seriously, every day? President Corona!

Who made me see a psychologist so I could work on my lifelong depression? President Corona again!

I can't help myself from repeating: "Thank You, President Corona!"

That's why I decided to write a book celebrating the positive impact brought by this tiny virus, which has probably saved my life.

Contents

Prologue

In the 7th century, an isolated lagoon became the shelter for desperate people. These people were forced to flee their homes by barbaric hordes, capable of barbaric acts. Their choice was between a horrible death or moving into a humid lagoon for the rest of their lives. Those people accepted the new reality and moved to the lagoon.

They learned to live with restrictions. The first years were harsh, but these constraints sparked the creativity of these fugitives. With time, they learned how to make the lagoon habitable. They even built houses, churches, and palaces on the water. They called their new home Venetia, nowadays known as Venezia or Venice in English, the most unique and magical city in the world.

1400 years later, Venice saw the amazing outcome of harsh restrictions once again. A new kind of wave forced modern humans to abandon their habitual posts. This time around, not even the lagoon was spared by this invisible force. The initial shock was soon replaced with wonder when for the first time in centuries, the waters of Venice were crystal clear. The waters became quiet as well, so quiet that even dolphins paid a visit to the city, replacing the noisy human tourists for once!

Humans were locked while animals took over city centers. Pollution went to all-time lows, and the air became breathable again. People were released from their duties and finally found time to spend with their families, educate themselves, meditate, draw, sing, learn a new language, work out, cook or simply rest. The world found time to rest.

The world was desperate for rest! It took someone special to stop the world, someone who has nothing to lose. Someone who has a clear agenda and doesn't stop before anyone or anything. Someone who doesn't fear consequences because doesn't have family, friends, fans, or allies to care about. Someone who doesn't care about what you think and doesn't need your permission or your vote. Someone complex, yet so simple and direct in taking action. Someone who doesn't compromise, because always has the upper hand. Someone you can deal with, but not negotiate. Someone who doesn't need the courage to take drastic measures, because to have courage you need fear, but this someone has no fear at all.

This someone is the first-ever President of the World! As such, this president has a unique set of skills and qualities. But what's more interesting are the qualities this leader doesn't have: compassion, disdain, empathy, sadism, ambition, vanity, charisma, and presence. This president doesn't feel either pain nor pleasure, love nor hate. This president doesn't, because he doesn't need to. Because he is not someone, but something... This president is the virus that drove humanity nuts, forcing us to reinvent our lives and to ask questions we had never wanted to ask. The first step to real progress comes with a question. The president posed us so many questions that we can only say: Thank You, President Corona!

My Personal Great Reset

What Was Going On?

Some day in January 2020

What did I want to do with my life? When I was spinning this question around my depressed mind, my phone rang. I hate phones ringing. They immediately make me swear and this time was no exception. I swore and then – a second later – I calmly said: "Hello?"

It was my friend Valeria, but that voice was not hers. It didn't even seem like a voice at all. It sounded like Darth Vader after having saved Luke from the Emperor! Only after a few seconds, those broken breaths managed to turn into words: "Ale... Al... Please! Come here! I c... I can't... I can't breathe!"

It was something serious! What the hell was going on with Valeria? At first, I tried to advise her to call an ambulance as her house is closer to the hospital than mine. Then I realized that in Malta, ambulances are never in a hurry. If you die, you die. There's no rush my friend!

I turned on my graffiti car and raced to Valeria as fast as I could, while I also called the ambulance (just to increase the chances).

The ambulance and I arrived almost at the same time, but I still followed Valeria to the hospital. I thought it was some leftover of her wild parties, but Valeria swore she hadn't partied since NYE.

I didn't have any explanation, nor I had the energy to find one. Apparently, not even the doctors were able to assess the reason for this sudden lack of breath in a 27 year old girl, who's perfectly healthy, fit, and vegetarian.

The doctors tried with a dose of Tamiflu or something similar, but to no avail. Valeria's conditions improved only when she was administered with pure oxygen.

She was dismissed after 3-4 hours, the time of additional exams, all of which turned negative. She was OK by then, except for muscle pain in the thoracic area, another mystery.

February 2020

I had just come back from an intense trip to London. The Brexit saga was finally over. The UK was free from the EU and the EU was free from the UK.

I was wondering what the media would have come up with to keep people entertained.

Everyone bet that the new hit would have been a nuclear attack from North Korea. Some even went as far as to predict a Russian escalation in East Ukraine, where rebels of the Donbass region were fighting regular Ukrainian forces to establish the independence of the region.

I hate predicting the future and hate even more who claims to predict it. These folks usually shoot in the dark and when they guess it right (big numbers' law), they twist their reading to make it look like they are prophets who have always known everything. If these folks can see the future so well, why aren't they filthy rich?

In retrospect, everyone can predict the future, once the future becomes the present...

Today, we see these folks bragging that they had seen the pandemic coming, or even that it was "all planned". However, I can't remember, nor could I find any posts or articles discussing this plan before 2020.

Everything related to a possible pandemic, up to 2019, was more suitable for a science fiction movie rather than a realistic outbreak.

Because, when you think about pandemics, you imagine crops of dead bodies all around, you imagine militarized quarantine zones, you imagine soldiers chasing and gunning down who breaks the rules, while some weird pathogen is turning folks into monsters ready to attack whoever is not sick yet!

Nobody thinks about the opportunities that a global-hitting event like the COVID-19 pandemic could carry along.

Like everyone, in those first days of 2020, I was skeptical. I quickly dismissed the new coronavirus from Wuhan. It was just the new sensation of the mass media.

I was confident my life would have been the same shit for years to come, until I would have figured out what I wanted to do with it. London had asked me what I wanted to do with my life. The only thing I knew is that I had to leave Malta. I was wasting my existence there. Malta is a wonderful place, but it can't be forever, not if you want to grow up. It was time to burn the Peter Pan's clothes, even because every particle of Peter Pan I had in me had already died since a while...

6th March 2020

It was another Friday at work. It was still the old world order. The new order was preparing to take over.

My last task of the week was to read the last email from the HR department.

"Please, take a moment to read this message. We want you to know that we are taking this situation very seriously. We will keep monitoring the developments and act accordingly to all the instructions coming from the local authorities..."

Well, once read the email, I packed my stuff, emptied my locker, then I looked at my desk neighbor, Sabine: "It seems like I won't come back to the office for a while. I don't care if the office will be staying open. It's time to try to work from home."

Sabine was perplexed, but I could see she was having the same idea. Our idea was right. Just one week more and our entire staff was ordered to work from home.

A strict procedure was put in place for those who needed to retrieve their belongings from their desks.

That day was the last time I ever saw the PokerStars office.

What did I want to do with my life? I wanted to change... everything!

1st April 2020

It was around 2:00 am when my flatmate Karolina knocked on my door: "Alessandro! Alessandro! Please! Wake up!"

What can possibly go on at that freaking time??? Is this an April's Fool???

When I opened, I saw the face of a really worried woman, scared I'd even say: "James is having another heart attack! Can you drive him to the hospital?!"

I immediately remember what every virologist was saying: "The coronavirus is mostly found in hospitals. Avoid hospitals at all costs!"

Yes, Sir... Well, if I had to meet the coronavirus, it was better if it happened in an heroic fashion, while I was trying to save a friend's life! James was upstairs barely able to speak. He sounded as if Darth Vader was force choking him! I quickly put him in my graffiti car and again, I raced towards the hospital.

Unlike the time before, this round the roads were clear.

Already after 10 minutes, the Hospital Mater Dei of Malta was on the horizon. Its majestic incinerators stood tall like gigantic Christmas trees, offering their red lights to the dark and cloudy sky.

In that dark situation, the bright side was the ER parking lot. It was completely empty for the first time in history! Perfect to drop my friend right at the entrance.

What was eerie though, was the ambulance parked there. It had its blue lights fully on, spinning around as if they had gone crazy. The vehicle was surrounded by four people in hazmat suits. It really looked like a science fiction movie. I was already wondering where the zombies were!

The hamzat guys stood there staring at us. That uncomfortable tension was broken by two nurses who came out of the building. They waved at us to come in.

As we approached the entrance, the nurses abruptly commanded to stop. One at a time, they started asking the following questions:

Have you been to China in the last 14 days?

Have you had any flu-like symptoms lately?

Have you felt any sense of tiredness, nausea, short breath or any musculoskeletal pain?

I replied "No" to all, while covering my mouth with my jacket. The masked nurses approached the termometer to my head. The movement felt as if they were handling a gun. In my mind I started panicking for 5 good seconds: "I know! Now they will detect fever! They will detect fever!!!"

The termometer said I was clear. The nurses stuck a yellow stamp on my jacket, certifying that I was safe and that I could walk freely in the premises! Yay! James passed the screening too, specifying right away that he wasn't there for the coronavirus, but for an heart attack.

We were left alone in the waiting room. The only other soul there was a food courier, lying on the chairs with his head broken. I wondered how he could have had an accident of that kind with no cars around.

After the empty parking lot, another miracle happened; James was called in for the check-up right after 5 minutes! 30 minutes more, and we were already home!

I liked that mess! When people stay at home, minding their fucking business, things start working smoothly! That's how you handle society!

Thank You, President Corona!

Why "Thank You, President Corona!"?

When I say the coronavirus improved my life, people look at me as if I were from another planet. They think I'm ironic or even that I have some mental issue. This is because we tend to believe that everyone reacts the same way. What's good for me must be good for the others too and what's bad for me must be bad for all the rest of us. That's where intolerance starts from. For centuries, people imposed their religion, their beliefs, their values, their ideologies on others because they truly believed that it was the best for everyone.

Reality is very different and whatever happens will always harm someone and benefit someone else. With his drastic policies, President Corona created a lot of trouble to many, but at the same time, he offered new massive opportunities to some others. I don't like to toot my horn usually, but in this case, I'm proud to tell how I did everything to place myself in favor of President Corona's agenda and benefit from it.

The way to become a President's fan wasn't straightforward. I had some setbacks too, but unlike you guys, I've never tried to look for a scapegoat for this situation, not even for a second. You are free to believe it or not. It all started from how I looked at the emerging pandemic. I started from the motto "Hard times make strong men"! I wanted to be a man of success and only strong men reach success! I knew that I wasn't tough enough though. I needed time to become a tough man. The problem is that I never had time to work on myself before the pandemic. Social life in Malta is very hectic. There is always some event to attend on this tiny island. There are always friends to meet, any day of the week. It's like being on a never ending holiday! That sounds great, but with time it can be exhausting, especially when in depth, you are an introvert. I needed to rest and take a long time for myself. Probably, this is the aspect I most thank President Corona for.

He came at the right time, giving me the precious chance of

taking a break from my intense social life.

My friends will never believe this, but I have always struggled with social anxiety, propelled by an ever present low self-esteem that made me suffer even from depression. Over the years, I learned to live with low self-esteem till the point that I even turned it into a strength point, so to speak.

Don't get me wrong, I wish I had a high self-esteem because there is no way that low self-esteem can be an advantage in life. You can only try to turn it into your favor as little as you can and make those tears be worth something. Without a low self-esteem probably I would have never approached the pandemic like I did. I would have never had the humility of trying to understand the dynamics behind the restrictions, and the reasons behind certain controversial decisions. I would have thought that the world moves around me and so I would have never accepted the situation. Thanks to my low self-esteem, I rather accepted that I know nothing, that I am no one and that I'm not smart enough to understand everything from a few clues. With a high self-esteem I'd have been roaming and roaring on social media, insulting who didn't understand "what I know" and who couldn't see "the bigger picture!". I'd have labeled people either "irresponsible" for "being selfish" or "naive" to believe what "they say" and criticizing every measure with the arrogance of having a "simple solution" for every complex problem.

In other words, I feel like a low self-esteem encouraged me to get smarter, kinder and intellectually more honest.

This constant feeling of never being good enough always pushed me to seek improvement in every aspect of my life, even though I knew that I could have never reached the desired level. Just the simple journey to an unreachable destination made my life a bit more purposeful. I felt stupid, unhealthy, ignorant, unattractive, weak, shallow, boring, poor, and under-skilled. With this mindset you either collapse for good or you try to elevate yourself.

I believe this background transformed me into a normal

"special" person, an individual without special qualities, but who managed to get out of mass mediocrity. I am not a genius, but I still have the strength and courage to think with my own head, the patience to look at an issue from different angles, and the tolerance to listen to the reasons of every party involved and to consider everyone innocent until proven otherwise.

I am not a scientist, but I always try to grasp as much as I can about science, always within the rigor of the scientific method, whose ultimate demand is to have the courage to accept uncertainty.

I am not a politician, but I still try to walk in the political class' shoes. By doing this, I learned to feel their doubts when called to make different groups, people, theories, needs and demands to coexist even when they are divergent. Politics is like a puzzle where you can't hope to frame all the pieces: some will always have to stay out, happy or not.

Now that you know how I think, you also know that it was natural for me to exploit a negative event like a pandemic to my favor. Having the world stopped before my eyes looked like the opportunity of my life! Since February 2020, I wondered: “How can I profit from this?”

Does it sound horrible? Maybe, but with profit I don't mean just financial gains. I wondered how I could improve myself, where to invest my new free time on, which bad habits to get rid of, which activities to prioritize and what projects to set aside for good. How to reset my course for the better, using this unique chance to correct what was wrong with my life as a whole! The pandemic marked my Personal Great Reset!

Financial Gains

The first aspect I tried to milk out of the pandemic was financial, I admit it. I've always dreamed of achieving financial independence, but I realized that I hadn't made any significant step toward it. For example, it was years that I didn't have a stocks account. It was time to open one. Armed with patience during my lonely evenings, I spotted the stocks I wanted to buy. In a market where everything was down, the plenty of choice was the real obstacle. Potentially, every name listed on an exchange could be a good buy, but I knew that not all of them, even the biggest firms, could survive the pandemic. I decided to set some criteria:

- The company had to offer essential services and to be still working,
- The company had to be in good shape before the crisis,
- The company had to be stranger to the pharma industry,
- The company's stock had to record the highest loss possible.

These 4 criteria made me invest mostly in banks, postal services and food or beverage producers. All of these sectors were essential, never halted and were in good shape as per criteria 1 and 2. All of these companies were not directly involved in the big vaccine race which turned the pharma industry into the biggest roulette in history. Spotting the winner of the COVID-19 vaccine race was just like horse betting, at the beginning of 2020.

After proper research, the industries that satisfied all of my 4 criteria were banking and postage. Despite these are essential services, banking and postal services suffered the most, but nevertheless they rebound quickly. My first easy capital gains in years! Though this was nothing... In the second part of 2020, President Corona raised my most precious asset from the dead. From the $8,000 area, Bitcoin rose brighter than ever, ending 2020 close to the $30,000 threshold! The pandemic marked the great comeback of cryptocurrencies. The classic actors of the financial sector had declared crypto dead for good. Now, they are all crazy about crypto assets, launching them into the mainstream realm! Thank You President Corona!

Muscle Gains

Though my financial situation improved significantly during the pandemic, my muscles got the biggest gains. At first I thought that the President bugged my plans, but in reality he pushed me to do more physical exercise! Ironically, the first time I heard about the coronavirus I was at the gym. I was lifting some (light) weights, right in front of a TV tuned on BBC News. The headlines talked about the expansion of this "Wuhan virus" with the first cases outside of China. It felt like such a distant threat, the nth exaggeration by the mass media, always ready to sell fear. I didn't imagine that a few weeks later I'd have to drop the gym!

My mind adjusted immediately to the new situation. I forbid myself to look for excuses, to get depressed, to even complain!

"There are billions of people who suffer more than I do, all year around. I can cope with a few restrictions!"

This mindset urged me to find an alternative to the gym. I had to start working out at home. This became much easier than expected, when PokerStars asked us to work from home! At work I'm obsessed with time optimization. That's how I came up with the idea of replacing my meaningless cappuccino breaks with push-ups breaks! Every two hours, it was time to stretch my legs, have a walk and then go down to the floor for my reps! I started following fitness dedicated YouTube channels. Chris Heria, Austin Dunham and Igor Voitenko became my mentors! I started from the fundamentals, discovering the right way to do push-ups. This brought me back to step one. From 20 wrong push-ups a set, I receded to 5 real push-ups a set, then 10, then even 20!

With Chris Heria I started to follow complete workouts, making work muscles I even didn't know I had! For the first time in my life I could see the features of my six-pack.

Working from home gave me the opportunity to change my diet for the better. I was cooking my own food, choosing personally the ingredients I wanted. As per my mentors' advice, all flour products were eliminated from my diet. This wasn't a linear process as giving up on pasta wasn't easy. Pasta is so luring, because it's so easy and fast to prepare and it makes you instantly full. In reality, pasta makes your brain full, but your body is still in need of proteins. What your body needs is the sauce you put on pasta. I came up with ideas of transitioning from pasta to a fully healthy diet. Lentils lent me a precious hand. Lentils are delicious! They are a boost of fiber and proteins that can even play the pasta role. Try pesto or Bolognese sauce on boiled lentils and you will be amazed! It will taste even better than pasta itself!

Cooking my own food was only one of the wonderful benefits of working from home. As I have already said, this new fashion allowed me to work my body much more than before. This led me to craving for more sport. This desire made me discover boxing! I've never got so many endorphins like when I practiced this sport in Spinola Bay, Malta. The scenario there is perfect, right in front of the sea, surrounded by plenty of luzzu (the traditional Maltese boats) and swimmers who look at you with curiosity. Everyone is willing to help you improve your technique and motivates you to give your best. The Scottish coach is always ready to drain every single sweat drop from you. It's hard as hell! There is no rest between the exercises, unless you call the plank position "rest"... But after the training is done, you feel like in heaven! All that suffering is rewarded with a boost in your self-confidence, with the awareness of having improved your health and right with it, the indescribable sense of feeling good!

Music

I love singing. Ironically, singing was the first hobby that pushed me to work out. How on earth? What the hell does singing have to do with exercising? Let me explain... Throughout 2019, I had been taking private vocal lessons. I wanted to perform better at the karaoke bar, but I also wanted to be able to record my own songs without destroying them.

On my last lesson, I still struggled to have volume in my singing. My voice was still weak, because my diaphragm was weak. My teacher suggested me to do hold the plank position while singing! That was the most genius exercise I had ever heard! I started practicing the same day she told me. It didn't take long before I could enjoy the challenge. Soon, I was no longer content with just a plank. Since I was down on the floor, I could go for push-ups too! A little of work-out got part of my daily life, until I decided to hit the gym again.

With new singing skills and plenty of time, I decided that music had to come back in my life! After a home-made album called *Lost in Malta*, for years I stopped publishing music until I had even stopped writing music all along.

When the pandemic came, I said to myself: "This is the right occasion! Now you have the time and even the inspiration to compose something!"

That's how I started to put notes together. When I came up with a riff that I liked, I started putting on words about the virus, until I baked *We Are the Virus!*
Recording this track taught me how to publish a song on Spotify. I refreshed my musical skills that were rusted after years of inactivity. Neurons are like muscles; you need to train them constantly or they will lose tone...

To be honest I wasn't happy with the final result. My voice sounded bad again, but at least this time it wasn't bad because of my intonation, but because of the scarce quality of my microphone. This pushed me to buy a new professional mic.

This was the sign that music came back into my life to stay!

My Social Great Reset

I couldn't imagine that spending all this time at home would have been so beneficial for me. This storm came at the right moment. As I said at the beginning of this chapter, I was tired of all the social life I had. Social anxiety was gripping me again. This anxiety has always been there, latent for years, but a particular event made it reemerge in all of its scary power. It came out from a deep hole after years, during my first ever trip to London, between January and February 2020. I didn't still imagine that President Corona was going to begin his mandate!

In those days, the world was hooked by the finalization of Brexit, and scared by North Korea's nuclear tests. COVID-19 was still referred to simply as "the coronavirus" and the biggest concern from the authorities was to know if you had been in Wuhan in the last 14 days.

My goal in London was to combine leisure, history and business. Leisure because I have friends living there and the fun offered by London is indisputable. History because Brexit was finally happening, and I wanted to witness it. Business because the biggest exhibition for my industry (ICE London) was happening during those days.

All the stars were aligned, it looked like this was going to be the trip of my life! However, ICE is what broke my confidence again. I went to know more about the industry I work in, just to realize that I knew nothing and that I had no networking skills.

My friend Alex was there too. To warm up each other, I pretended to be a customer interested in his company's services and using my experience in the payments industry, I posed him some nasty questions. It was the right and funniest way to break the "ICE" and get into business!

I was keen to explore, observe, interact while also learning as much as possible. Learning what? I had no idea, but I felt something extraordinary was due to happen! Shortly after, I bumped into my former manager Carla. I followed her for quite a few meters while she was talking with her colleague, until I broke into the conversation from behind and said with a low tone voice: "I agree!".

The laugh lasted very short: "Yeah, sorry we're heading to a meeting now! See you later!". Then I met John, my cameraman buddy, but he was as busy as hell filming around the venue. When I was getting bored, I found Dimitrios! He's a salesman promoting slots. He was in a good mood as usual, very welcoming and warm! Like all the others, Dimitrios was busy with meetings, but we agreed to catch up after the conference to "hit and destroy London!".

The more I was walking through the stands, the more I felt useless. "What am I doing here?".

Having a lot of friends there was a boomerang too. They were all busy with meetings. Comparison was inevitable. My mind started to shoot me as if it had a machine gun: "They are cool, I'm not. They have a serious role in iGaming, I don't. They are smart, I'm not. They have a purpose, I don't. They are sociable, I'm not" and so these overlaid voices in my head put me down and knocked me out.

Depressed was my state in the middle of day 1 at ICE. That's how I felt when I received a text from Elena, a girl I met at the AIBC Summit in Malta, where I had a stand for Aliencoin in the start-up village. That was a networking experience totally opposite to what ICE was until then. The AIBC 2019 (Artificial Intelligence and Blockchain Summit) boosted my confidence! It gave me hope that I can network with professionals, that I can be convincing and entertaining, that I can sustain certain kinds of conversations and that possibly, I can be a true businessman and even a salesman!

But in London I was feeling like the AIBC was an illusion. The AIBC was in my home place after all and it regarded a specific sector, still brand new, that nobody understood clearly yet and that I got to know before the others: the blockchain. I felt like the ICE was slapping my face to wake me up and bring me back to a harsh reality.

"Well, maybe I can help you connect within the iGaming world" is what I texted back to Elena. We met at Costa Coffee. Her hug lifted me up. She was with her business partner Anna. They are self-employed copywriters specialized mostly in the blockchain sector, but with the desire to expand in the iGaming field. They asked me how to move. My suggestion was to hit the payment processors, some of which were trying to propose crypto payments to the online gambling business. They seemed to agree, but first they needed a coffee and to discuss some business between them. I left the girls with their meeting and I didn't see them again... And again, I felt useless...

Of course, the rest of my trip in London got better when I met my friends living there, but it didn't erase the mark of the ICE. This anxious state of not being enough for the people I interact with came again to prominence in every social interaction I was having after my London trip. A night at the bar, a gathering at some friend's or even more a date were no longer pleasant moments but authentic exams. I needed a break from everyone but I was afraid of saying no to my friends' invitations. The President gave me a strong argument to stay home and avoid social interactions for the time I needed. I could focus on myself as described before

I could elevate my soul and feel more valuable each day. It was like the social distancing gave me time to turn from a bug into a butterfly. When the summer kicked off, I returned to social life more confident than ever, energized, recharged and happier! I went through a series of magical parties, doing crazy stuff, trying new activities, meeting new great people without ever feeling the anxiety of being judged, tested or screened. I was myself again and people liked me! And not having tourists in the middle made me enjoy the Maltese summer like never before! Everything was easily accessible, activities were queue-less, the roads were free from traffic jams, and the air was finally breathable! For the first time I could see the stars from my balcony and I broke my personal record of shooting stars seen in one night! All of this wouldn't have been possible without President Corona!

Of course, not everything was easy, and depression hit me again on several occasions after the summer loop faded. From August 2020, even in Malta we had to face some restrictions and even here the President started to spread a certain paranoia. The mindset I used to face the pandemic was toward balance. I must be careful, especially in not infecting the vulnerable, but at the same time I must be careful at not closing enough to let my mental health fall apart. Since I am someone with depressive tendencies, I couldn't afford to let myself go.

My philosophy was that if the President won't catch me, I'll get out of the pandemic stronger, wealthier and healthier than ever. If the President will catch me then he will put an end to the very suffering of living. Because living in the end is a fatigue. Yes, life can be wonderful, but it's also expensive and not only financially. I know this is not the most inspiring line in this book, but that's what I feel about life at the moment of writing. Probably that's why I accepted the kind offer PokerStars made to all of its employees: a free therapy session "to cope with these tough times". For me these times were not tough, but they were still a great chance for a final resolution to my innermost problems. The perfect ground for a final battle against my innermost demons. Thanks to PokerStars, I had the pleasure to know a therapist. I will call her Dr Blue. She's a very sensitive, acute and caring therapist who helped me look at my inner child. Only by taking care of my inner child, I can finally break that wall of insecurities built over many difficult years.

Practicing Self-love

Depression comes from lack of self-love. Self-love is a wide concept. Self-love is about being kind and caring with ourselves as much as we would do with others. During my first therapy session with Dr Blue, we commented how normally we are so kind and supportive toward others, but how poorly we treat ourselves. It happens more often than we realize. How many times have you had someone making a mistake and the first thing you did was to comfort them, telling them: "It can happen! Next time you'll be more careful. We learn through mistakes!".

How many times when YOU did something wrong you started yelling at yourself: "Shit! I'm just a fucking moron! I ALWAYS do like this! I'll NEVER do anything good in my stupid life!!!"?

The first homework Dr Blue gave me was to repeat kind words and practice self-love meditation. It is something simple in theory. You just have to focus on your breath and repeat simple statements like "I'm good enough. I'm special. I'm amazing. I'm attractive. I'm charismatic" and any kind of thing you want to feel like. In practice for me, this was harder than expected. Saying that "I'm special" or even more that "I'm attractive" sounded so stupid that my brain took a huge effort to even formulate such statements! It was like going to the gym for the first time! My therapist told me this is normal. Our brain needs time and effort to create new connections that favor new ways of thinking and new thoughts. You need to train your brain to develop new connections if you want to change your mind, just like you need to train your muscles to create new mass if you want to change your body.

If you want to go through this same kind of journey, I recommend you to visit a therapist, even via Zoom if you are afraid of President Corona. If you can't afford one, there is still great material available online. About self-love in particular, I recommend the guided meditations by celebrity therapist

Marisa Peer. Her voice is hypnotic and indeed she masters hypnotherapy. She even has an online course in this discipline. Of course, I don't agree with every aspect of Peer's methods. For example, I don't agree when she tells you to lie to your mind "cause your mind doesn't know that you're lying" and other possible shortcuts. I don't believe in shortcuts and I know that a good psychotherapy is a long process, not a "rapid transformation". But Peer's career revolves around the entertainment industry, where who you are is not really important, but who you look like is what matters. Trick your mind until you believe to be the character you want to be. As always, nothing and no one is 100% good or 100% bad. Check the percentage and then take from them what's good for you.

Working on My Mental Health

As I said, thanks to PokerStars I met a therapist. I was so anxious the first time we met, but I must admit that being it online made it easy. I felt less embarrassed and not having to cover a physical distance to Dr Blue's studio reduced the time to accumulate anxiety.

The experience with Dr Blue was brief but intense. She advised me on how to handle my resignation from PokerStars, how to communicate this to my friends, but mostly to my parents. I was so afraid of their disapproval and I was sure they would have gotten desperate, they would have tried to infect me with their fears. They tried indeed, but they didn't succeed. I felt relieved and also excited that finally I could travel, after 1.5 years trapped on a tiny island. Freedom was waiting for me!

I must confess that my first idea was to travel to India, to volunteer in the tragic fight against the Delta variant. You may think Delta was just a joke from the media, but at Stars I had Indian colleagues, and I can't forget the burden that situation had on them. Some of their closest ones were among those poor people left in the streets, without oxygen to breath. It is something I don't wish for my worst enemy.

For a while I searched for volunteering programs to India, but my attempts were unsuccessful. The few I managed to find demanded skilled staff. Of course, they needed badass folks used to these extreme situations.

As I realized I couldn't help others, I kept on helping myself. That's how I started to travel around Europe, until the stars aligned to transfer me to Lisbon. Never would I have imagined to settle in Portugal! It happened by chance. One day, my friend Valeria (yes, the one who got without breath in January 2020), told me about her trip to Portugal. She fell in love with the place. She told me about its beauty, its welcoming and kind people, its tasty food, and its affordability. Under Valeria's pressure, I added Portugal to the final stages of my itinerary.

Just when I was in Portugal, I was interviewing for a blockchain company. I had been traveling for 6 months. My spirit was satisfied. I was ready to work again. Everything fell into place. By February 2022, I was part of the SwissBorg family. SwissBorg is an app to help people manage their crypto wealth. In a nutshell, you can use SwissBorg to buy and sell crypto using your bank account, and you can even park in the app to generate passive income out of your crypto assets, in a fashion similar to what wealthy people do with traditional assets.

It was great, but yet, I still felt empty. On top of it, moving to a new city where I didn't know anyone, with a full remote job, meant to be lonely. And I knew that my depression was more alive than ever. Restarting my life from scratch meant to be lonely, but it also meant to have a clear path to rebuild it from its very foundations! I was more determined than ever to embark in a serious therapy journey.

I dedicated a whole day to find the best therapist in Lisbon and boom, the healing process began — and it continues to this day! I will call this therapist Dr Red.

Most of you might think I'm crazy, or that psychologists are just a bunch of nerds duping you with pretentious analysis that means everything and nothing. Maybe you imagine them as coaches telling you what to do. Nothing like that. The work of the therapist is not to give you orders, nor to tell you what to do with your life. The work of the therapist consists in detecting what sparks psychological suffering in you, where it comes from, and how to overcome it, so you can finally fulfill the life you dream of. No life can be fulfilling without peace of mind. Mental health is as important as physical health.

In order to shed more light around a topic that is still covered by a fog of misconceptions, I want to share with you what I had learned in one year of therapy. The pivotal points about therapy are that:

- It is scary
- It is painful
- It is expensive
- It can be frustrating
- It requires commitment
- It requires trust

It is scary

Let's be honest. Telling a stranger things that you've never told anyone else, and you would never tell the closest ones to you, is damn scary! Allowing this stranger to analyse your mind, the very core of your being, of who you are, is an insane act of courage. You place your entire being in a position of absurd vulnerability. What if this person will use this info against me? They could destroy me! Yeah, I confess that was my first thought.

There are other legit fears, like: What if the therapist is not good? What if I'm a hopeless case? What if I find out I'm not the one I thought? Maybe I'm a psycho? A violent person? Do I have some secret perversion? Do I really hate my family? Am I autistic? Am I gay without knowing it? Am I schizophrenic?

The list can be as long as all the things you can be afraid of being.

Though the initial mistrust and fear surrounding the act of exposing yourself can sound daunting, I feel that there is a heavy load of misconceptions building up this fear. When you mention "psychological therapy" to people, they immediately connect the expression to something like: "psycho", "madness", "violence", "the Joker", "Arkham Asylum", "Hannibal Lecter". When the connections are not so grotesque, they fall into stigma, like: "weak", "pussy", "crying baby", "spineless", "snowflake", with specific connotations according to the patient's gender. If you're a woman, you are "a hysterical girl, a spoiled princess who's never happy with her life", while if you're a man, you are "a weak boy, who cannot control his emotions because he never grew up as a real man."

It is painful

If you're into macho masculinity, then fear not, my friend! Because therapy is all about pain, and real men don't avoid pain, they endure it. What if I told you that going to therapy is the most macho thing you can do nowadays?

And if you're a girl? Time to pull out your feminism and show that girls can do what boys do! I'm sorry boys and girls, but you're both humans after all, and pain is an indiscernible component of human existence. You can even be non-binary, gender fluid, or hermaphrodite; there is no gender out of pain's reach.

If I have to be honest, I could describe psychological therapy as a surgery for the mind. The problem is that this surgery is carried out when you're fully awake...

Where did I find the motivation to continue? Because a pain of such scale brings handsome rewards! A big pain once, in order to stop suffering a little bit every day for the rest of your days. Take it as an investment. Would you rather live with an annoying wisdom tooth for all of your life, or would you go for a few seconds of excruciating pain to have it removed once and for all?

It is expensive

Yes, therapy is expensive, especially if you go for the best, like I did with Dr Red. It is as expensive as your latest smartphone, your new car, your branded clothes, or your weekly sushi dinner. Expenses are a matter of priority. Before wondering if therapy is expensive, wonder what you need the most. If you really feel down, it might be worth carrying the same car for a couple of years more in order to fix yourself first. Your priority #1 should be yourself, not your car, not your image, not your social status. If you believe therapy is expensive, wait and see how expensive your life will be without it...

If you're really struggling financially, then the pandemic created a massive opportunity: online therapy. Online apps helping you find a counselor for everyone's pocket are popping out like mushrooms and some platforms even offer financial aid! The first of this kind was BetterHelp.com for North America, but there are other countless options for each nation and region. Just google "online therapy" and you will find something fit for you. Some even offer a first free session, so you have nothing to lose. You just ran out of excuses here, uh?

It can be frustrating

How long did I take to see tangible results in my life? I'd say 5-6 months of therapy, keeping an average close to 1 session per week, and my mood was still very susceptible. I'd say my mood stabilized only after 9-10 months into therapy.

There were moments when I thought that I had started therapy too late. I remember one evening in particular. I was walking on a bridge, crossing a railway in Lisbon while it was raining. My mind looped into a scene from *The Killing Joke*, one of the most compelling and creepiest stories in the Batman's universe. Just when the final match between Batman and his nemesis, the Joker, seems to reach the climax, something weird happens.

Just when the Joker seems powerless, he asks Batman: "Well, what are you waiting for? Kick the hell out of me..." But Batman's reply is still, firm: "No! Not this time..."

Gotham's protector states that the Joker is not alone! He can rely on Batsy's help! This seems the offer of a lifetime for the Joker! But you know, when someone is doomed, maybe he's doomed forever... Does a violent reaction erupt from it? No, no... Something more extraordinary happens, given the character. The Joker simply sobs. For the first time, he loses that threatening smile and with a lower voice utters: "Thanks, but no... It's too late for that... It's far too late..."

When you lose hope, it doesn't mean hope is no longer there. Open your eyes!

It requires commitment

People can't change if they don't want to change. A therapist is not a shaman who hypnotizes you, rewires your brain with magic formulas and wakes you up by snapping the fingers. Therapy demands your will to navigate yourself out of muddy waters. Do you want to believe free will exists? Good! Then, you must also accept the big responsibility that comes with it. If you want you can. If you don't want to, then don't blame the therapist. What if the therapist is bad? Give the professional at least 3 sessions, then follow your guts. Don't look for results (those might take even months to show up), rather focus on your reactions.

- Do you feel listened to and understood?
- Do the therapist's explanations make sense to you?
- Can you relate to what the therapist tells you?
- Can you feel challenged without feeling insulted?
- And most of all: Do you trust your therapist?

It requires trust

Trust is the spine of every therapy journey. No real healing can occur if you don't open up with your therapist and don't allow him/her to work for you.

Trust takes time. Don't expect to click right away. Even when you feel confident you're sharing everything, most likely your subconscious is still holding back on key elements. Give your therapist time, and give yourself time too.

And what if there isn't enough trust? If you cannot open up on a specific issue, you may want to face other easier issues and go back to the darkest one later, when trust will be stronger. If a lack of trust surrounds every topic, then it might mean you picked the wrong therapist. It's nobody's fault. Just move on and start again. Therapy is a process, and you must take into account setbacks of this kind.

If you're not convinced yet, that's fine, hopefully it means you don't need and that your mental health is fine.

If you feel the need for therapy but you are still scared, the only thing I can tell you is to surround yourself with people who will support you and protect you. I've been lucky. I must admit that I hesitated and postponed my decision more than once.

I think I got lucky with three people who encouraged me to put my pieces together. Person #1 was assertive in telling me that it was time to start a serious therapy journey. This person was talking from above her experience and was leading me by example.

Person #2 called me after a really bad depressive episode I had, so bad that she started crying and out of those tears she felt compelled to tell me straight: "You are sick! You must do something! It is the wake-up call! You can no longer postpone it or the uphill might get so steep that that climbing it will become impossible!"

And Person #3 gave me the final push. It was a gentle touch, as unexpected as it was her visit to Lisbon. She asked me over a glass of wine: "Have you ever been close to die?"

I replied: "Well, I can't remember if I've ever been close to die out of an accident. But... willing to die and getting close to do it... Oh yeah!"

Her soothing smile disappeared, replaced by a worried expression. Her voice became deeper and inquisitive: "Are you still willing to die....?"

Me: "Yes... I am."

At this point she said something as unexpected as her visit to Lisbon: "If you need to talk, please call me, but most importantly, consider therapy. For many this comes with a stigma, especially for men, but there's nothing wrong with it. Therapy is not an act of weakness. It's actually an act of strength and courage."

Some human beings have the gift of making you willing to be a better person.

Curing My Social Media Addiction

My personal Great Reset inevitably involved my use of social media. At the beginning, President Corona pushed me to use and interact on Facebook even more than I used to do. In 2020 I reached levels of pure addiction. After all, COVID-19 was the big topic and I couldn't resist sharing my opinions (or bullshits) just like everyone else! I noticed that even at work I checked my phone every 5-10 minutes. I even started a pervert self-rewarding cycle: "OK, when you finish this mini-task, you can check Messenger."

It was a destructive pattern, because the reward should be something more beneficial, like a walk, playing with the dog, eating a nice snack or doing more abs crunches, for example. My constant use of Facebook was consuming my time and draining my energy. On top of this, seeing the ignorance and arrogance of people regarding the COVID-19 pandemic, twisted my nerves every day. Yes, I managed to channel that rage in a creative way by jotting down these very pages, but it wasn't enough. I felt the duty to reply to every stupid comment that disputed the hard work of all those trying to manage the emergency, going down even to explain to these folks what was really going on and why their theories didn't make sense. I think I even convinced some of them to at least review their beliefs but for the most part, it was a pointless exchange of insults.

After all, every social media platform is designed to be addictive. Notifications are always displayed in red, creating a sense of urgency that is hard to resist. You always crave to check if your crush likes your post. You always want the upper hand in every futile discussion. You want to share that hilarious meme with your best buddy. All of this is fun, until it's too late.

The statistics showing social media addiction are greater than any gambling site, by far. Just consider that social media users are estimated to be 3.1 billion. That's more than the adult half of the world population!

Of these 3.1 billion, 2.9 billion are already mobile users, accessing their accounts on a daily basis. 210 million people are thought to use social media in an addictive, compulsive way.[1]

Facebook was definitely taking too much time from my work and my personal life. When even depression knocked the door and the first draft of this book began to get too dark, I decided it was time to take a break from Facebook. Overnight, I logged off from Facebook and uninstalled Messenger.

The only social media platform I allowed myself to use (and not abuse) was LinkedIn, the only one I really need. LinkedIn is the only social media where you can share valuable content, make precious connections, have fruitful discussions (most of the time), find jobs and above all, learn something new every day! While people on Facebook comment in underwear, on

---

[1] https://mediakix.com/blog/social-media-addiction-statistics/

LinkedIn they comment in suit and tie. If I ever have to become an influencer, I want to become a LinkedIn influencer!

It's on LinkedIn that I posted the daily progress of my personal challenge. In February 2021, I decided I had to stay completely out of Facebook and Messenger for at least 15 days. I had never been off for so long since I signed up with Zuckerberg back in 2009! That was at the peak of the last great financial crisis... What a coincidence!

Here you can read the chrono story of my abstinence from Facebook. Original typos included, because you know, typing from a phone sucks.

Life without Facebook: Day 1

I just logged off Facebook and Messenger and I intend to stay out for at least 15 days. let's see how it goes...

Life without Facebook: Day 2

I feel lighter. yes, a couple of times I grabbed my phone tapping where Messenger used to be, but over all, I feel better. the day has been way more productive both at work and off of it.

Life Without Facebook: Day 3

I still have the compulsive need to check the phone every 10 minutes. Checking for something that is not there... However, I noticed that the few chats left via WhatsApp or LinkedIn are more meaningful and productive!

Life Without Facebook: Day 4

Getting much better! Like magic, I received two important calls on WhatsApp (nobody ever calls me) and one of these changed a person's destiny for the better! She found a job!

Life Without Facebook: Day 5

I don't think about it anymore. As a consequence, I came back to some projects that had left behind. I can't disclose more, but I can tell you that Aliencoin is getting back on track! Stay tuned!

Life Without Facebook: Day 6

it was a though Monday, tougher than usual. not because it was Monday, but because I didn't sleep. while discussing Aliencoin and watching the game, I drank 4 cappuccino: result that I didn't sleep and felt dizzy and dehydrated all day long. luckily I couldn't indulge on Facebook and waster further energy

Life Without Facebook: Day 7

It's been a week now, but it feels like a month. So far I feel more energy and I see I have more time for my stuff!

Life Without Facebook: Day 8

All good but I before going to sleep I bumped into some anti-vaxxers that touched my nerves. I let myself go to some acid comment. Taking Facebook bad habits on LinkedIn is not good. A setback compared to the other days...

Life Without Facebook: Day 9

Half way of this journey is behind. I see I'm still tempted to flame for a fight. I had the "wise" idea of commenting Richard Branson taking the vaccine, and ended up insulting anti-vaxxers and conspiracy theorists.

That generated good content for my upcoming book Thank You President Corona! But on the other hand denotes that I'm taking bad Facebook habits into LinkedIn and this must be fixed...

Life Without Facebook: Day 10

I still feel an increase in energy and I engage more with my friends. It was an unforgettable day from a social point of view, and by "social" I mean real human interactions!

The only shade came by a thought which crossed my mind for a second: I wondered what will I find once I will log in back to Facebook on Day 15....

Life Without Facebook: Day 11 and 12

A strange Sunday where I got a late evening nap that powered my night writing. Creating a book during the night is magical, you feel your ideas coming together, it's like dreaming while writing!

No wonder that few hours later, a very productive and satisfactiory Monday began. I've never felt so happy to start working on Monday!
Thank You President Corona!

Life Without Facebook: Day 13 and 14

Not much to say and this is good! it means that not being on Facebook is no longer a bug deal for me. If I have to signal something is that among the increased overall energy I feel now a will to work that I haven't had in months!

Life Without Facebook: Day 15

And here we are! it was easier than I expected! I wanted to see if I could live without Facebook and Messenger and yes, I can! Not only, I can live even better! I'll discuss the benefits I noticed in detail during the upcoming days. Let's see what I'll find once I'll log in again, though for tonight I'll procrastinate this step...

I posted this every day, because I wanted to inspire other people to follow suit. I believe that there are more people addicted to social media than cigarettes or gambling and by far.

I want to raise awareness around this, because I don't want a world where you cannot live without social media. I even dream of a world where you can live without the internet if you want to. The so-called "right to be disconnected". I firmly believe that something must be done before it's too late. After all, we already had this massive effort to protect who's vulnerable to COVID-19. Yes, COVID-19 is perceived as an immediate risk, while social media addiction doesn't fill up our hospitals altogether. It doesn't because harm from social media is not direct and so it's not easily measurable. We can assume that out of those 210 million people addicted to social media worldwide, many are at risk of suicide. It takes a little to get tired of living. Even if we estimate a conservative 1%, we can have 2.1 million of people dying of suicide induced by social media abuse. Those are more or less the same deaths from COVID-19 during 2020! And you can be sure that in the years to come, these sad numbers are set to increase. Many teenagers may not be able to turn 18.

Back to my personal experience, the major benefit of breaking social media addiction is that you get your time and your energy back, rapidly. To use Marisa Peer's words, this is a very "rapid transformation" and the funniest part is that I procrastinated my come-back on Facebook by 17th of March. Not because of Saint Patrick's, but because I needed Facebook to help the NGO Friends of Kenya Malta.

I didn't feel any rush to log in back. I had never been so proud of procrastinating something!

Thank You President Corona!

Finding a Purpose

I had never really known what my life's purpose was. What am I doing in this stupid world? What's the meaning of life, if the world where it takes place is full of hate and violence? What's the meaning of life, if the only real thing it offers is death? What's the purpose of life, if you live in a world you don't like?

I don't like this world. It's too twisted. Even when I manage to switch from these negative thoughts and start to enjoy life, I still feel guilty for all the beings who suffer, now, in this very second. Why must it be like that?

In 2020, I felt like I was not important for anyone. I didn't feel important for anything either. I saw my job as useless, for the company, for my colleagues, for our players and of course, for the society as a whole. I felt I didn't deserve such a great company like PokerStars and such talented work buddies. I wasn't making any difference. I was not having any impact. For sure, I wasn't making the world a better place. I felt stuck in a loop, while I was wasting my youth playing the watchdog under the title of Quality Assurance Specialist.

I felt useless for my friends too. I lived with the conception that I didn't deserve the friendship of such fantastic people. Of course, all of this was just in my head, but in a depressive mental state, there's no way I could acknowledge this as a fog of misconceptions.

Then one day, President Corona came and he told me to write this book. I had a mission. I had to inspire people on how to turn an era-defining, negative event for the world, into an era-defining, positive event for their lives. I felt I could teach people how to turn the tables to their favor and change their lives for the better. I could teach them to see the good where the others cannot. I could help them understand science and politics. I felt I could guide them on how to recognize conspiracy charlatans. I could challenge people to think! All I had to do was to write a book. I finally remembered I had always wanted to write a book! I could be useful! I had finally found my purpose! I was keen on living again! Yes, I can say out loud that President Corona saved my life, literally... Thank You, President Corona!

How President Corona Changed the Economy

The IMF issued catastrophic predictions of a possible financial crisis of the same proportions seen during the 1929 Great Depression! Media didn't hesitate a second to amplify these dim views and turn them into a granted truth, an inevitable dire destiny. However, reality turned out to be very different. During the Great Depression, up to 15% of the global GDP was burnt and the consequences lasted until 1932. It took a bloody World War to lead the economy back into prosperity.

The 2009 Global Financial Crisis burnt 5.57% of the global GDP, which grew again right in 2010. President Corona burnt just 2.77% in 2020, contracting the global GDP from $87.39 trillion in 2019, to $84.97 trillion in 2020. For 2022, the GDP is set to exceed $100 trillion for the first time in history! Not numbers worth of an apocalypse...[2]

---

[2] https://www.statista.com/statistics/268750/global-gross-domestic-product-gdp/

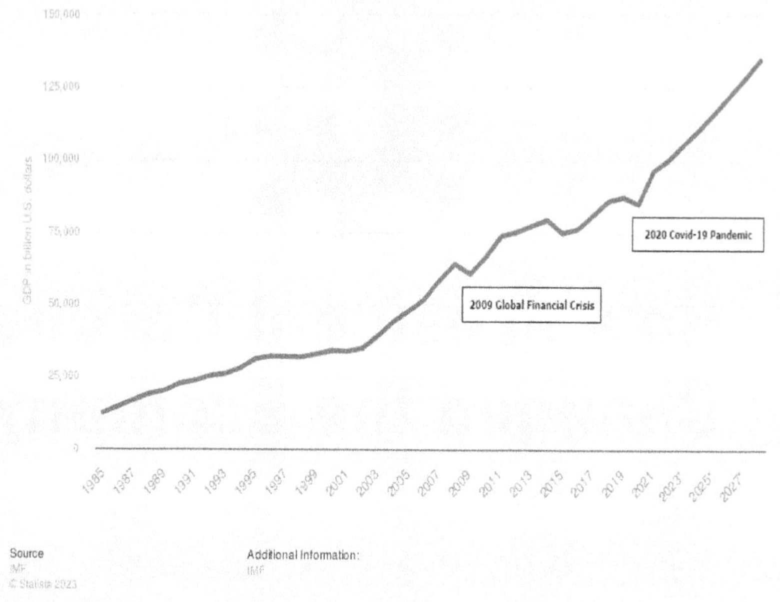

This graph shows once for all that all the bad rumors about President Corona's destructive effect on the economy were fabrications, designed to tarnish his reputation. They've mistaken President Corona's shocking reforms with those financial crises provoked by human insatiable greed.

Did you know that financial crises are cyclical events which occur at precise intervals, always following the same pattern? The major financial crises, like the one from 2008, start from the real estate market. When the real estate market crashes, everything is sucked into the abyss. Real estate market crashes are very predictable. The pattern was spotted by Fred E. Foldvary, an economist and professor at Santa Clara University.

In his extensive study *The Real Estate Cycle and the Depression of 2008*, covering 2 centuries of real estate market, he created the following table:

Peaks in Land Value	interval in (years)	Peaks in Construction	interval in (years)	Peaks in Business	interval in (years)
1818	--	-	--	1819	--
1836	18	1836	--	1837	18
1854	18	1856	20	1857	20
1872	18	1871	15	1873	16
1890	18	1892	21	1893	20
1907	17	1909	17	1918	25
1925	18	1925	16	1929	11
1973	48	1972	47	1973	44
1979	6	1978	6	1980	7
1989	10	1986	8	1990	10
2006	17	2006	20	2008	18

As you can see, the real estate market in the US crashes on average every 18 years. This interval can be shortened or stretched by special events, like a World War for example. Hyperinflation and energy crises can shorten it, like happened between 1973 and 1979.

The American real estate market is traditionally the most liquid in the world and it's heavily linked to the American financial world, so entrenched with the economies of all the globe. It's obvious when the American real estate market crashes, the entire world economy suffers with it. If you predict this, you will predict everything.

If we look at the above table, we may deduct the next land value peak that was set to be for 2024. The question is: has President Corona accelerated the peak or has he postponed the date of the next real estate crash? There is not a certain answer. The high inflation hitting in the aftermath of the pandemic would suggest the market is going to crash before expected, but on the other hand, we saw how the economy is on steroids since the lockdown season is over. I don't like "predicting" events. I hate it. This is material for fortune tellers, which I disdain, deeply. This is your call. Do your own research and then extract a conclusion out of it, after having collected evidence, not before!

If you're interested in real estate, I highly recommend you an amazing course on LinkedIn Learning, *Pre-investing: Before Investing in Real Estate* by Symon He. It contains valuable information and explores sides of the real estate market you ignored. You will love it!

President Corona requested a conspicuous death toll and many years of life from our brothers and sisters. Was this huge sacrifice rewarded by the first president of the world, somehow?

Well, if you were an industrialist, the benefits for you would have been instant. According to a detailed report by Swiss bank UBS (so someone well informed), the world's wealthiest guys had their wealth climb 27.5%, up to $10.2 trillion, from April to July of 2020! Even the total number of billionaires existing in the world reached a high of 2,189, new record from the 2,158 of 2017.[3]

You are not surprised by these numbers, are you? You smelled this plandemic was inflated to stuff the elite's coffins. However, what might surprise you is that these billionaires are not so involved in Big Pharma. They mostly come from the heavy industry or the tech industry. We all know Jeff Bezos topped an astonishing $200 billion wealth during 2020. Amazon billed more than any other company, due to people relying more and more on e-commerce, while stuck at home. However, fewer know the new richest person of China is not from tech, nor from the heavy industry. His name is Zhong Shanshan, founder of Nongfu Spring, a packaging company which bottles nothing more than water, the most precious thing there can be! Water brought Mr Zhong to own assets worth $58.7 billion.[4]

Mr. Zhong has also a participation in the pharma industry, with a 20% stake in Beijing Wantai Biological Pharmacy

---

[3] https://www.bbc.com/news/business-54446285

[4] https://www.bbc.com/news/business-54276902

Enterprise, which skyrocketed when it was listed on the stock exchange during 2020. The firm was even working on a candidate vaccine for Covid-19 but failed to deliver one. However, the company has been active in the development of Covid-19 antigen test kits (rapid tests). Tests are surely more profitable than any vaccine. You don't need to invest much to develop a test, while you need to invest enormous resources to develop a vaccine, without the guarantee your money will ever come back. A vaccine is supposed to be used only twice on each person, while antigen tests can be used multiple times over the same subject, making the product more profitable. Even including boosters, test kits will always be much more profitable and a safer bet than vaccines. The vaccine race reminds me of the golden rush in the Far West. Who made money weren't the gold diggers. Most of them found nothing or found death, just like most of the pharma companies that tried to come up first with a Covid vaccine. Who really made easy money in total safety were the pick sellers. You need a pick to find gold! You need a test kit to find COVID-19!

Tests were crucial to limit the use of lockdowns, or at least, to make sense of these harsh restrictions. You say that lockdowns ruin the economy. I even heard some saying that the economy is more important than health!!! So, let me ask you: why do you want to make money? What's the ultimate purpose of it? Maybe you forgot it, but I guess you started to work because you needed money to do certain things: eating, drinking, having a shelter, having fun, reproduce, traveling, educating yourself (though the latter doesn't apply to most of you, I know) but... Could you really fulfill all of these goals without health? Not really, right? So, can you tell me what will you do with your money if you won't have good health and if, in the worst scenario, you will die?

China proved how the faster you act, the earlier you recover production. In the democratic West, the motherland of human rights, everyone thought about their own interests, delaying the necessary measures, hoping to limit the economic impact. In China everyone worked for the greater good and we could see the results. This is why most of the Chinese are happy with their dictatorship. Democracy is simply not for them and I respect that. Let's not forget nowadays democracy rhymes with idiocracy, with individualism, with the mantra "I'm more important than anyone else".

Job Losses

How many jobs were actually lost during the pandemic? According to the International Labour Organization (ILO), 114 million jobs were burnt by the pandemic. Mostly were lost due to the disruptions brough by lockdowns or travel restrictions which slowed down supply chains. If we count the overall hours of work erased during 2020, those make up to 255 million full-time jobs.[5]

Even who didn't lose their jobs saw their working hours reduced, which impacted their finances. These numbers are impressive. To have a better idea, we just need to look at the jobs lost worldwide in the aftermath of the 2008 recession. On that occasion, "just" 8.4 million jobs were lost during the downtrend started by the subprime crack.[6]

The good news is that while the 2008 recession was caused by a structural crisis of the global financial system, with long-lasting effects everywhere, the pandemic was caused by an exceptional event bound to make society stronger. The labor market was shaken. It got a severe fever, but because of this, it has now acquired immunity against future crises of this kind. Future pandemics won't disrupt the labor market like COVID-19 did. And unlike in 2008, many of those 114 million people could start a new job soon after the restrictions were lifted.

[5] http://www.ilo.org/global/about-the-ilo/newsroom/news/WCMS_766949/lang--en/index.htm

[6] https://www.ilo.org/wcmsp5/groups/public/---ed_emp/---emp_policy/documents/publication/wcms_174964.pdf

President Corona brought many opportunities which we will check in great detail later.

The massive loss of jobs represented an unexpected opportunity for some migrants. Going back home was not as terrible as imagined for some people, especially for migrants coming from East Europe. Traditionally, Eastern Europeans looked at West Europe as the Eldorado, the land of opportunities and insane salaries to support their families in the motherland. This is still true. Salaries in the West are still 3-4 times the salaries in the East. The difference is that the cost of living in the West is becoming unbearable, particularly in the capital cities. Maybe it's not so worth living in Paris when the rent takes away more than 1/3 of your salary[7].

Meanwhile, West Europe's growth has been slow if not stalling for the last decade. States like Poland, Romania or Czechia on the other hand, recorded growth levels worth of China. Poland, with its strong manufacturing industry propelling the GDP growth, was nicknamed "the China of Europe". Since their entry in the EU, Eastern states grew at a steady pace, while also improving the living conditions for their citizens and in many examples, they managed to eradicate a corruption so widespread during the years following the collapse of communism. When I was in Poland, I was delighted by how foreign investors were happy with the absence of corruption in the country, at least at public administration level. If you want to open a business, you can do so easily and you are ready to start in less than a week, without any presents for friends of friends. All is done in total transparency. The catch is that many Eastern expats were unaware of the big changes their states underwent during their absence. When President Corona deported them back to their countries of origin, many startled. A Lithuanian friend told me she was shocked: "When I left, crime was high, and men were violent alcoholics. Now Vilnius is the nicest city

[7] Convensional wisdom says that rent should never take more than 1/3 of your salary.

ever! I'm staying here!"

Job opportunities are on the rise. Poland leads the new manufacturing industries, but Warsaw is also attracting top service companies, even stealing businesses from post-Brexit London. Warsaw is one of the few European cities with skyscrapers. These high buildings proudly show the logos of EY, Volkswagen, HSBC and other giants, in front of a luxury Marriott Hotel. The Baltics are leading the digital revolution which is taking the financial sector by storm. Most of the fintech startups dwell either in Vilnius, Riga or Tallinn. The latter is the place with the highest startups per capita, in the world. Silicon Valley is nothing compared to Tallinn and the cost of living is 10 times lower, for now at least... You may want to consider a real estate investment in the Estonian capital.

The low cost of labor in the East is not the only reason why investments are going more and more into this area. This is one of the few areas in the world where you can have professionalism, high quality education, integrity and loyalty for reasonable prices. Bureaucracy is slim and efficient, making room also for a flexible and dynamic labor market, which offers the right compromise between employer and employee rights. Safety is another winning factor. Contrary to the stereotypes, Eastern Europe is safe, even safer than the West. If in Silicon Valley it is not unlikely that a crazy dude will break into your offices with a machine gun, in Tallinn the chances are nearly 0.

Why are the Baltics so good in IT? The reason lies in their aggressive Russian neighbor. Russians never really understood why the Baltics left the Soviet Union and didn't continue in the post-communist world under the new Russian Federation. They didn't understand why the Baltics didn't want to be part of something greater, instead of defending their small territory and their niche languages. Obviously, this lack of understanding has always been insulting for the Baltics, who worked hard to defend the independence required in 1991. With the rise of widespread computer technology, some members of the Russian minorities attempted to destabilize the Baltics governments with aggressive cyber-attacks, the most serious of which occurred in 2007, when a Russian hacker managed to put down the websites of several banks and institutions in Estonia, through a brave DDOS attack to the banks' servers. From that moment, Estonia invested massively in training its population in cyber-security. When you excel in cyber-security, you can do whatever you want in IT. Instead of weakening the Baltics, Russia made them stronger and more skilled than ever, paving the way to the Baltic Valley!

Now, President Corona is accelerating the shift of power from West to East Europe. These states are getting their brains back for the first time in years. They are taking talents back to their homes, to the countries which educated them.[8]

I wouldn't be surprised if in 20 years the situation will be the opposite of what we have seen so far. Polish, Czech, Estonians happy in their home countries with highly paid jobs, hiring maids, masons and plumbers from Germany, France or Italy. It will look weird for us, but normal for the future generations.

Something opposite happened to who was already living in their own country. After losing their jobs, some people realized there could be more opportunities abroad. A new kind of "high-end" migration intensified during the pandemic. People from rich regions like West Europe and North America, started to move to Portugal, the new hot spot for digital nomads, professionals and pensioners. Portugal offers many opportunities, but mostly it offers an unparalleled quality of life, tasty food, variety of nature (ocean, rivers, forests, mountains), plenty of history and very kind people, most of whom speak English fluently. Portugal is becoming the new home for many Westerners disappointed with their own motherlands. In Portugal I met especially Californians, frustrated for the most diverse reasons. Who was fleeing from the Trump administration, who complained about the outdated

---

[8] https://www.economist.com/europe/2021/01/28/how-the-pandemic-reversed-old-migration-patterns-in-europe

infrastructures, who could no longer stand the prohibitive cost of living, who was escaping from the wildfires of California and who was a doctor and 2 years spent in the ICU trying to save stubborn anti-vaxxers fanatics wasn't worth anymore.

Disrupting Holidays

The pandemic had a dramatic impact on our holidays. President Corona wanted to remind us of the real meaning of our holidays. Think about the word: holi-day > Holy Day > Sacred Day. They were born to bring up a deep meaning. They weren't conceived as a feast of unleashed consumerism. I always laugh at those Christians who observe the vigil before Christmas by eating lavish and fancy seafood dishes. The vigil was conceived as an occasion of repent. Fish was usually consumed because it was cheap in ancient times, being abundant in the ancient seas. Nowadays, a real vigil would better be observed with chicken, than an expensive sea bass...

However, nowadays everything must be monetized, nothing is sacred anymore. So, let's see what the impact on the major holidays observed by humans was.

Christmas

2020 was the year when Christmas never happened. In most countries, people were restricted from seeing their families. Especially for those who live abroad, this was more than a simple inconvenience. When you live far from your hometown, you might happen to see your close ones only once a year, during Christmas. In December 2020, travelling was impossible. In some states, it was even forbidden to move from town to town! The emotional cost was immense, something you can't quantify. Canceling Christmas was the cruelest act by President Corona. For the first time, I allow you to think Fuck You, President Corona! You have permission!

President Corona left no room for the Christmas markets, no

chance to make small trips, no space for the queues waiting for the latest electronic gadget. People always complain that Christmas has become a mere show in honor of consumerism, turning the holiest moment of the year into the most profane. These people are right under many perspectives, so they should be grateful that President Corona invited us to reconsider the real meaning of Christmas. We had the chance to think more about how to spend Christmas with our loved ones rather than to check what present to give them. Consumerism made us forget that the most important Christmas present is just Us. When everything will be back to normal, think how you can spend more quality time with your family members, your true friends, your loved ones. Take extra attention to the oldest ones, as they are not supposed to be here forever. For the elderly, time is the most precious asset. The President made it brutally clear. Don't think you can make up for your absence with some expensive gift sent via post. A simple pack of chocolate delivered in person will be much more appreciated. Oh please, don't rush away the minute after you gave them the gift. Stay for a chat at least. Even half an hour will be enough to turn a gift into a meaningful present... moment!

In spite of everything, holiday sales kept growing in most parts of the world, even with shopping malls deserted or limited in capacity. People simply turned to online stores to fill their voids. It was reported that Christmas holiday expenses grew by 3.6% compared to 2019.

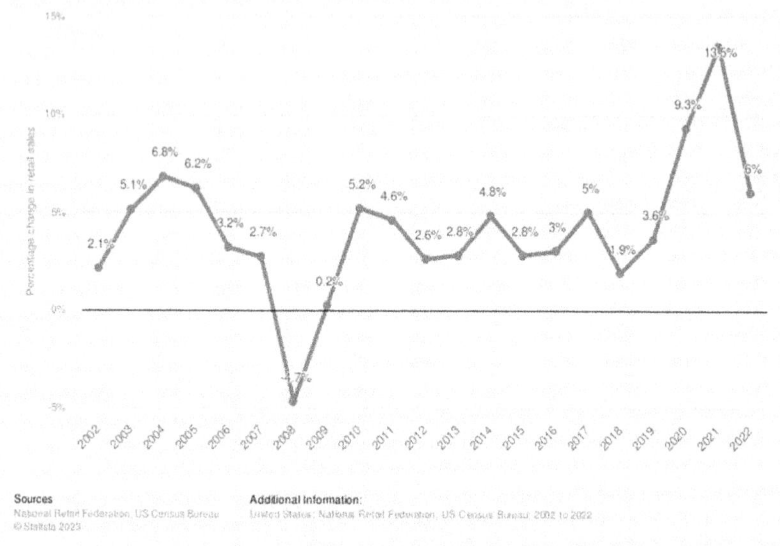

Year-over-year growth of Christmas holiday retail sales in the United States from 2002 to 2022[9]

The pandemic seems not to be the worst economic catastrophe ever. The 2008 financial crisis had a way worse impact on sales. 2008 was the only year in the 21st century when Americans contracted their Christmas expenses. There was no money at all back then. Companies were winded up overnight. This time around, companies were simply forced into an unexpected holiday, but ready to start again as soon as the viral storm was gone.

[9] https://www.statista.com/statistics/209299/year-to-year-percentage-of-change-in-holiday-retail-sales-in-the-us-since-2000/

Diwali

Christmas was not the only holy festivity affected by the pandemic. Diwali is the Indian "Christmas", the Festival of Lights, which celebrates the victory of light over darkness, good over evil and knowledge over ignorance. This sounds like the expertise of scientists that overcomes the ignorance of people. It can be the laif motive of President Corona's campaign. Diwali is celebrated mostly by the Hindu people, over 5 days that may fall between mid-October and mid-November, according to the year. In 2020, Diwali started on Saturday, November 14.

Like Christmas in the West, even Diwali 2020 saw an increase in sales. The Confederation of All India Traders (CAIT) even reported record sales for the Indian traders. The main factors seem that in the months preceding Diwali, people spent only on essentials, avoiding any other unnecessary expense, in a time of uncertainty. This allowed Indians to approach the holiday with considerable firepower. Another important factor is that CAIT called for a massive boycott on Chinese products, following the latest tensions between India and China over an area called the Line of Actual Control. The Sino-Indian border is disputed by the two powers since 1959. For the first time in 61 years, Indian and Chinese troops escalated tensions, though both governments deny having authorized their soldiers to get loose in a melee. Whether the provocation was sought or not, Indian traders decided to react by banning as many Chinese products as possible from their shelves. The estimated outcome was a Diwali record turnover of 72 billion Rupee (about $9.8 billion) and 40 billion Rupee ($5.4 billion) of losses for China. Great shot! This is beautiful, because it shows that an economy without China can exist and can even thrive.

If the West really believes the Dragon is evil, it should start to think seriously about alternative partners. China is not the only country capable of producing cheap merchandise. You have countless alternatives nowadays, including India itself.[10]

Islamic Rituals

As you can expect, a religion full of rigorous rituals like Islam was highly impacted by COVID-19. The impact was all visible in one of the most common Muslim rituals: the Friday prayer. This is the only prayer that is required to happen within a mosque. All the other prayers might be performed at home, without any requirement to be physically present at the mosque. For a good Muslim, it is essential to attend the sermon on Friday noon. This particular prayer is called *Congregational Prayer*. In this ritual, believers listen to the sermon, while their shoulders touch each other. As you can understand, social distancing is not guaranteed. Congregational prayers would be the perfect vehicle for the virus to spread. Friday prayers were never canceled since 622 AD. Not even wars or previous epidemics were capable of discouraging believers from attending the Friday sermons, shoulder by shoulder. As you can imagine, authorities met a stiff resistance to prayer bans.

Some Muslims tend to be totally reliant on God and see the pandemic as a God's tool to punish humanity for its latest sins, like pollution, climate change, social inequality, erosion of customs and all the excesses of modern society. No wonder there is a current within Islam that would suggest giving up to President Corona and let him decide who has to go, without trying to step in his way. Nevertheless, the very prophet Muhammad warns Muslims to take care of their health.

He even talks specifically about outbreaks:

"If you hear of an outbreak of plague in a land, do not enter

[10] https://zeenews.india.com/economy/diwali-sales-up-10-8-goods-worth-rs-72000-crore-sold-amid-boycott-of-chinese-items-2325078.html

it; if the plague outbreaks out in a place while you are in it, do not leave that place."

I'm sorry for the Muslims denying the efficacy of the anti-Covid measures. Your prophet approves them 100%...[11]

Indeed, Islam is all about preventing infectious diseases. We can just look at the emphasis the Quran puts on hygiene. Here you find procedures on how to clean yourself, commands to wash regularly your hands, to wash your private parts after your physiological needs and to bath regularly, all of these written in a time when hygiene was a joke and more than one millennium before Robert Koch formulated the basic principles of the germ theory of disease. Even Muhammad is said to have enunciated: *"God loves those who are clean"* (2:222).

Beside the Friday prayers, Muslims were asked to give up another important part of their daily lives: family. In Islam, the concept of family is wider than what we have in the West. It is not limited to the partner, children and parents, but it extends to grandparents, grandchildren, cousins, brothers and sisters in law. It is a requirement to take care of your entire family. That's why Muslims tend to live together in big numbers, or if they are in a developed and rich country, they still tend to live in the immediate vicinity. The lockdown measures prevented Muslims from being in direct contact with their wider families, something that surely had a toll on their mental health. Not living up to your own values can cause much stress and let an uneasy sense of guilt grow within you, even when you are not at fault at all.

Muslims were even prevented, in most jurisdictions, from burying their dead according to their burial ritual. Forced cremation became the norm and this led to some harsh protests, especially in South Asia. This is the eternal fight between secular state and religion. Burying your loved ones according to what your God says is important, but this ritual

[11] https://theconversation.com/how-coronavirus-challenges-muslims-faith-and-changes-their-lives-133925

cannot pose a risk to the others.[12]

Much of Islamic tradition requires mass social contacts. Hajj is the greatest example. Hajj is one of the biggest pilgrimages in the world, capable of attracting more than 2 million pilgrims to Mecca each year. Hajj is the famous pilgrim to Mecca that every adult Muslim is required to fulfill at least once in a lifetime, provided that the person has the financial means to undertake the journey. It gives life to that huge human swarm walking around the black cube in Mecca. As it is one of the five mandatory pillars of Islam, cancelling Hajj could be a big problem for the faithful. Despite the apparent shock, this holy mass ritual was cancelled about 40 times in Islam history. The reasons for previous cancellations were mostly wars, political crises, but even outbreaks, like a cholera outbreak in 1846 or a mysterious plague brought from India in 1831. This time, the main concern was the economic impact of cancelling the Hajj. In 2019, Hajj related tourism generated $12 billion, which is 7% of the Saudi annual GDP![13]

The COVID-19 pandemic posed also a political challenge to Saudi Arabia, in a world, the Islamic one, where politics and religion go hand in hand. On one side, the Saudi authorities had to protect public health. On the other hand, they had to upkeep to the expectations of millions of pilgrims around the globe. The solution brought an unexpected outcome, capable of surprising even the West. While deciding that only residents in Saudi Arabia (of all nationalities) could take part to Hajj in 2020, for the first time in history, Crown Prince Mohammed bin Salman decreed that women would be able to attend the ceremony without a male guardian (mehrem), as long as they travel in groups. This is a huge milestone for Muslim women, a huge leap toward their independence from men, a symbolic, yet significant loosening of men's control over women, which is still deeply rooted in the Muslim society. The surprising shift

[12] https://www.dawn.com/news/1609078

[13] https://www.trtworld.com/magazine/hajj-2020-the-economic-impact-of-the-saudi-ban-on-international-pilgrims-37520

met the favor of many husbands too. Men see this as an opportunity for their wives to fulfill their religious duty, while they can stay at home taking care of their kids. Family was a big obstacle for many Muslims. It can be hard for a couple to travel together, if they cannot afford a baby-sitter and/or they don't have close relatives able to help with the kids. This new rule made female presence in Hajj skyrocket, with an estimated female participation of 40% out of the only 10,000 pilgrims who made the final cut for the limited edition of 2020![14]

For sure Muslim women have to thank the progressive reform called by Prince Mohammed bin Salman, but perhaps this reform came with pressure from President Corona...

Art and Culture Matter

Art was the sector that suffered immediately because of the safety measures. They say that you don't eat with art and culture, nor you cure a disease. Nevertheless, without art and culture we have nothing.

If there is something which makes humans stand out from other animals, that's culture. Culture is the whole array of all the knowledge, art, technology, customs and social norms that shape and influence the values and ways of living of a determined group. It is the backbone of every complex society. Art in particular is the ultimate way of expression for a human being. It allows individuals to communicate to others not only their ideas, but also their feelings, emotions and their understanding of the world. Their artworks may influence others with the idea conveyed, leave them indifferent or even spark outrage. Whatever the reaction, art is essential in keeping a complex society alive. It is fuel for every highly developed brain.

Art appeared in human history from the very moment

[14] https://www.voanews.com/a/middle-east_mecca-women-take-part-hajj-guardian-rule-dropped/6208472.html

primitive men started to paint their caves. It marks the moment in which humans found extra time to think about concepts not directly linked to satisfy the basic physiological needs. They started to wonder about the reality around them. Imagination found a door into the human mind and art became the way to materialize such imagination into the physical world. Primitive paintings of animals soon developed into more complex representations. Painting got hand-in-hand with storytelling, the other basic form of culture, which is so hard-wired in us that even today we need it in every context. It's not a chance that most likable and sociable people are good storytellers. Imagination is like a liquid; it adapts to the container where it is stored. As the container expands with technological progress, so does art. The industrial revolution brought us cinematic art and reinvented the way of enjoying other forms of art. New themes began to form the backbone of the new tales.

Culture as a whole expanded exponentially and it became an industry itself. It soon became an industry so big and complex that even non-artist people became necessary to support the new scale: collectors, dealers, assistants, salespeople, agents and the kind of figure that we can't associate with art and creativity: lawyers! These professionals became key for art in the modern capitalistic world. Capitalism brought the new concept that you can privatize an idea and profit from it. The fact that you are the first one to conceptualize an idea, entitles you to possess that concept, just like the early American explorers were entitled to own the virgin lands they had discovered. Just like discovering a land, experimenting with new art ideas takes courage, especially if these ideas challenge the social status quo.

Censorship has always been the archenemy of art and culture, since the dawn of time. Authorities have always been afraid of new ideas that could challenge the status quo and push people to overthrow the current rulers. The great dictatorships of the 19th century had a strict control over art for this obvious reason. If you want to build a totalitarian regime, where everyone's head is filled with the same exact thoughts, you need firm control over art and culture, which are the primary means through which humans form their thoughts, values, habits and beliefs. Propaganda is the art of controlling art. Propaganda is not exclusive to dictatorial governments. Propaganda can be used even by democratic governments. A good example was the thought of Sir Keith Joseph, the right-hand of British Prime Minister Margaret Thatcher. When he was Secretary of State for Education and Science, Joseph envisioned an educational system focused on developing the skills required by the labor market, rather than educating kids on cultural and artistic aspects which may grow their critical thinking abilities. In this way, the new generation would be more easily controllable and more prone to keep on voting for the same party. A similar attitude is used in Poland, where the PiS government recently reformed not only the school system, but the very content of the text books so they can reflect the vision and values of the party in charge, like Euroscepticism, conservative values and religion, in the hopes of harvesting future loyal voters and reduce future political opposition.

Even the left is no stranger to censorship over art. The culture of political correctness often threatens to put down content that "might offend minorities" even when the content's nature is purely satirical. The newest attempts at extending the boundaries of hate speech crime, seem directly influence what a person can say or think. This might look legit on a first level. Spreading hate toward minorities is one of the most infamous acts you can do and like history teaches us, never leads to anything good. However, hate speech must not be confused with satire, nor with legit fears that citizens might develop toward a group due to anecdotal events. In the latter case, you don't solve the issue through control, punishment and fear, but through listening and understanding, even educating if necessary. Worrying examples of this trend are the latest laws in the UK or Germany, the motherland of political correctness. About Germany, I'd like to tell you a funny anecdote. Once I met a girl who streams on Twitch and also streams silly, satirical videos on YouTube. One of the videos she made mocked ISIS, the fearsome Islamist force which is ravaging the Middle East. It's a short video where she plays an improbable ISIS representative, answering questions from a journalist. The jokes are brilliant, a bit sharp but not edgy. At the end of the video, she even specifies she's Muslim herself. Few weeks after the upload on YouTube, German authorities "politely" asked her to put the video down, threatening that it will be impossible for them to grant her a residence permit in Germany...

But censorship is not the only danger for art and culture. Even the liberal, capitalistic world has its doses of poison. Even if not deliberately, a free market society can still harm art and culture as we mean them. When artists become too obsessed with pleasing their audience, they can be tempted to quit from actual art and start producing pure, empty entertainment. A clear, recent and easy example is the Star Wars franchise. George Lucas conceived it as an innovative, bold space opera (the first of its kind), full of action and entertainment of course, but with deeper meaning too, like the evil is more charming than good and how even the greatest of democracies can fall into the darkness of a ruthless empire. But since Disney took over the franchise, Star Wars movies lost any meaning and structure in their stories and all the focus nowadays is on action sequences, visual effects and new fancy characters to create toys and other merchandise from. Of course, George Lucas hoped for huge profits from Star Wars too and enjoyed them a lot, but it's also true that he liked to provide something more than entertainment with its creation. If you create remarkable artwork, mostly with your own resources, putting genuine passion in it, then profit is well deserved. There's nothing wrong with profiting from your artistic work. What is wrong and perverted is having profit as the sole goal of your "artistic" work. This obsession has grown each year in Hollywood, till the point that now it's impossible to see anything new, challenging, shocking or nearly original coming from LA.

It is obvious that an all-free-market economy can seriously damage art and culture, even pushing the best art on the brink of extinction. That's why most of the democratic countries have foundations to support independent artists. Funding and promoting alternative art have nothing to do with socialism or even worse communism, as someone speculated. Its purpose is rather to prevent that very cultural decay which may lead society towards dystopian scenarios. A non-profit drive funding of culture is essential to create a culturally diverse society. Why? The less diverse our culture is, the more we resemble sheep and the more we resemble sheep, the more we need firm guidance. A dystopic shepherd dog is the only possible outcome in the long term.

President Corona played a really hard test on artists and with them, on the whole society. How to support artists in a free market society if they cannot work?

"The culture sector, which employs more than 30 million people globally, has been hit much harder than expected by the coronavirus pandemic and its fallout, the UN Educational, Scientific and Cultural Organization (UNESCO) has said, urging targeted policies and actions to help it weather the crisis."

- UN News, 22 December 2020.[15]

The pandemic posed an enormous strain on governmental coffins. With most of the public money gone to directly manage the health crisis, few dimes were left to support art and culture during this difficult moment. President Corona called the artistic world to do what is best at: being creative. That's how many artists immediately searched for new, alternative ways to reach their audience, produce new kinds of crafts and organize festivals through unprecedented formats. Online

[15] https://news.un.org/en/story/2020/12/1080572

events and festivals became the norm as of March 2020. To describe this revolution at its best, it's worth mentioning the *Social Distancing Festival.*[16]

This cultural event was born because of the pandemic when Canadian playwright Nick Green was told that his upcoming musical *In Real Life* had to be canceled because of the new health threat. Green was also terrified by the idea of spending so much time alone, barricaded at home, when the lockdown hit Toronto. He had to do something, to keep art alive. These were the conditions which led Nick Green to set up a website hosting digital art from his fellow artists. That's how the *Social Distancing Festival* was born. The idea was so needed, that Green didn't have to do much to promote the event. He just shared it on his Facebook profile:

"The initial Facebook post — in 24 hours, it was shared about 1,700 times. So, obviously, the next day I was like, 'I think I better move on this.'"

- Mick Green, Social Distancing Festival founder[17]

You can imagine the feedback. In just four days, the festival had almost 80,000 unique visitors a day! The success was so overwhelming for a home-made platform, that Green had to ask help to keep up the site with this traffic magnitude.

The festival is colorful, well presented and with many powerful artworks which encapsulate all the most pressing issues of this period, from the very COVID-19 pandemic, to the struggles of the Black Lives Matter movement. You will find highly suggestive works in all forms: visual, audio and video.

I wish my friend Kristina had a similar success with her *Art Banquet* online exhibition. Her idea was very similar, though she chose a specific topic to open her virtual exhibition: discussing the new synergy between art and technology and

---

[16] https://www.socialdistancingfestival.com/

[17] https://www.pri.org/stories/2020-03-30/artists-flock-only-festival-still-during-covid-19

how we became reliant on the latter in every aspect of our lives.

Technology is what artists must thank the most. This book is called *Thank You, President Corona!*, but all the artists around the world must say: *"Thank You, Technology!"*

The Internet filled the void left by desert venues. It provided room for online exhibitions like what happened for Nick and Kristina. It was also the place to host concerts, cultural conferences and much more. The Internet brought even a new, unprecedented form of art, or better, it made digital art possible to trade like physical art. This was possible with a new technology using the internet: the blockchain. This digital oracle, powering the famous (or infamous) Bitcoin, gave birth to the so-called Non-fungible Tokens (NFTs). When you turn a digital file, let's say an image you created, into an NFT, this file will be inscribed into the blockchain and will get a unique digital code (hash). This code will help the blockchain to determine which file is the *original* and thus, it won't be possible to create another copy with the same hash. With this trick, that file will be unique, no matter how many other copies you will make. If this sounds nonsense, don't worry, you're not alone. This technology is so new that most people don't understand it at all. Whoever seems to understand it, most of the time has no clue at all. From a financial perspective, the definition given by Investopedia sheds some more light:

"Non-fungible tokens or NFTs are cryptographic assets on blockchain with unique identification codes and metadata that distinguish them from each other. Unlike cryptocurrencies, they cannot be traded or exchanged at equivalency. This differs from fungible tokens like cryptocurrencies, which are identical to each other and, therefore, can be used as a medium for commercial transactions."[18]

I know that you think that cryptocurrencies are a bluff, a

[18] https://www.investopedia.com/non-fungible-tokens-nft-5115211

bubble, a scam. In reality, they are none of that. They are the future and they are already the present too! I don't want to talk about cryptocurrencies here, or I should write another book. For me, it's enough to let you understand the disruptive force of the NFT technology and how this new tool was the salvation for many artists during 2020. Have you ever heard of Christie's? The big auction house, set in New York, which sells only the finest and most sought-after art pieces in the world. As recalled on Christie's website:

"In recent years, Christie's has achieved the world record price for an artwork at auction (Leonardo da Vinci's Salvador Mundi, 2017), for a single collection sale (the Collection of Peggy and David Rockefeller, 2018), and for a work by a living artist (Jeff Koons' Rabbit, 2019)."

These guys are smart and they are good at their job. You can guess it by the figures they are used to handle... So, what if I told you Christie's sold an NFT for... $69,346,250! Yes, you've read right! The artwork in question is called *EVERYDAYS: THE FIRST 5000 DAYS*, by graphic designer Beeple (aka Mike Winkelmann). This NFT is a collage of photos and illustrations made by the artist for 5000 days (13.5 years!), every day non-stop. It's the first all-digital artwork ever sold by the auction prestigious house, but it won't be the last...

NFTs are changing every art, including music. Kings of Leon released a limited edition of their 2021 album *When You See Yourself* as NFT. The sale of this album was hosted by YellowHeart, a new blockchain-based music platform.[19]

Fans could purchase two types of NFT: one would give you access to an exclusive vinyl copy of the album, while the other, a life-time ticket to all the band's future shows; a "Golden Ticket" to be their VIP fan, forever! The sales were open in January 2021, for two weeks and generated over $2 million for Kings of Leon, $500,000 of which the band donated to *Live Nation's Crew Nation* fund to support live music crews during

---

[19] https://yh.io/

the pandemic. The gesture is laudable and shows how creativity can overcome any obstacle![20] [21]

Another important artist to follow this trend was Grimes. Elon Musk's ex-wife couldn't hold back from riding the crypto revolution. In March 2021, Grimes sold almost $6 million worth of NFTs. They are mostly images and short videos about planet Earth and planet Mars, featuring her original music. She chose a platform called Nifty Gateway, but you can post your works also on other platforms like OpenSea, Rarible and even on the new Binance NFT marketplace. NFTs are in their early days, but the possibilities are already endless...

NFTs are reshaping sports as well. It may sound stupid, but there are people spending tons of money to purchase exclusive short videos released by the NBA, showcasing the best points of the weekend. As the videos are released as NFTs, they gain a great value in the eyes of the basketball fans. NBA Top Shots is the name of the NFT app, launched in October 2020 and capable of raising more than $120 million in the first 5 months of service![22]

The cinema industry is being invested by the crypto revolution as well. *Zero Contact*, starring his majesty Sir Anthony Hopkins, is the first full-length motion picture to be released as NFT. Even the way it was produced was innovative and so COVID-19 style. It was produced remotely, from 17 different countries, while the world was still experiencing restrictions. Zero Contact is going to open exciting scenarios for the cinematic industry, especially for the independent filmmakers.[23]

The upcoming documentary *Ethereum: The Infinite Garden* has been funded through the sale of NFTs on the blockchain

---

[20] https://www.nme.com/news/music/kings-of-leon-have-generated-2million-from-nft-sales-of-their-new-album-2899349

[21] https://www.theguardian.com/music/2021/mar/04/kings-of-leon-to-release-new-album-as-a-non-fungible-token

[22] https://nbatopshot.com/

[23] https://decrypt.co/80650/see-the-trailer-anthony-hopkins-nft-film-zero-contact

blogging platform Mirror, raising a total of 1,035.96 ETH, breaking the 750 ETH goal set for the campaign.[24]

Yes, independent filmmakers can already rely on crowdfunding hubs like IndieGoGo or Kickstarter to realize their dreams, but NFTs will help them raise their game to the next level, building a cult around their projects even before they are complete.

But before the cinematic industry could embrace and enjoy the NFT wind, it had to go through enormous sacrifices. Almost all film productions were halted in March 2020 and most of them didn't resume until summer. The number of times the new James Bond movie, *No Time To Die*, was postponed, became a meme. We can call it, *No Time To Screen!* But the most anticipated movie for me was *The Batman*. Halting this movie is one of the few things I hate President Corona for. Plus, he even killed a member of the production: on 31st March 2020, the dialect coach Andrew Jack died of COVID-19.

Jack wasn't the only excellent victim of the disease. On 16th April 2020, my artistic Olympus lost another hero: Luis Sepúlveda. He was the author of *Historia de una gaviota y del gato que le enseñó a volar* (Story of a seagull and the cat who taught her to fly). An outstanding, touching animated film was made from this 1996 book, with the title *Lucky and Zorba*. This movie caught my heart as a kid. It's one of the few movies which ever made me cry! At the moment of writing, I'm going to watch it again to pour additional tears. Consider also how long the agony was for Sepúlveda: diagnosed in February and died in April....

"Several countries have already issued orders for meticulous preservation of official records related to the pandemic. This not only underlines the gravity of the current situation, but also highlights the importance of memory

[24]https://ethereumfilm.mirror.xyz/3SV8gLXHIW8Ot45h3RL7aOgDINxN2hjLfFVOvyatB2A

institutions in providing the records or information management resources necessary for understanding, contextualizing and overcoming such crises in the future. At the same time, records of humanity's artistic and creative expressions, which form a vital part of our documentary heritage, are a source of social connectivity and resilience for communities worldwide, it is essential that we ensure that a complete record of the COVID-19 pandemic exists, so that we can prevent another outbreak of this nature or better manage the impact of such global events on society in the future."

– Moez Chakchouk

Shaking the Stock Markets

Whenever a crisis hits, the stock market plummets. On 9th March 2020, the Dow Jones fell by 7.79% in one day, setting the new negative record in the exchange's history. Three days later, the record was renewed at 9.99%.[25]

If Wall Street suffered massively, the other exchanges around the world didn't perform any better. London's FTSE 100 recorded a groundbreaking 31.1% loss for the month of March 2020![26]

On 9th March 2020, the Paris stock exchange, EuroNext, lost 10.53%, in just one single day!

The stock market is that annoying barometer supposed to tell you how the world is going. It is the first one to get excited when something good happens, and the first one to get depressed for even the simple collective feeling that something can go wrong! No wonder that the markets get unbearable when a new disaster begins.

People like you love to think that the stock market is a big theater operated by powerful puppet masters, who enjoy

---

[25] https://www.statista.com/chart/21082/worst-single-day-losses-of-the-dow-jones-in-21st-century/

[26] https://capital.com/ftse-100-news-in-march-2020

entertaining the world with their financial lies. It's true to some extent, but do you wanna bet that even this time you share responsibility?

OK, here is how it works. You like buying things, especially those you don't need. Let's say that you are buying an energy drink which we will call Blue Cat. This company performed really well in the last years and to raise more money, it decided to land on the stock exchange. How can Blue Cat make money out of the stock market? Simple; they divide the company in shares, they apply to be listed, they pass all the compliance and transparency requirements to get listed and finally the stock exchange approves their request. From then on, Blue Cat can sell their shares to a wider audience. Why don't Blue Cat sell their shares privately? Because in a private transaction usually you sell big portions of the company, in some cases even the majority stake, which means to sell the company and for a price per share lower than what you can get on the stock exchange. There, Blue Cat can sell some shares for a higher price and without losing control over the company. Being listed also means more exposure and free publicity for the brand! It gives you prestige. It proves your company is solid and established.

Blue Cat keeps on rising, so more and more people want to buy Blue Cat shares. Why? Because if our fluffy blue is increasing the profits, it means that its shareholders will get higher dividends, aka more money. Who doesn't want shares that issue good dividends?!

Over the first year of listing, Blue Cat shares increased by a wonderful 80%! This means, if in January 2019 one Blue Cat share was $10.00, in January 2020 it's worth $18.00! Without considering that the company paid $1.50 per share in dividends! Not bad! If you bought those shares at the beginning, you have a 30.00% annual ROI! You dirty speculator!

The cat is purring, but then President Corona comes into the game! People are worried, they hear that everything must be shut down for the common good. Businesses must halt. A lot of people will be locked down in their homes, which means some of them will be unable to work, maybe even unpaid for a whole month during this crisis. They will focus on buying only the essential goods to live, to even survive and an energy drink is not essential at all. Blue Cat will have lower revenues in 2020. It might even lose money! You can forget you will get any dividend from it. Your only hope is to sell your shares at a discount, hoping to find some big investor who has the liquidity to buy and sell only after the market recovers.

You may say, yes but those big investors are the elite which controls the world! Yes and no. There are a lot of people capable of living without working for a month and not all of them are super billionaires. It's natural that big investors like Warren Buffet earn a lot from these disruptions and it's no secret that speculators like George Soros can provoke a crisis on purpose to profit from it (see 2012 war in Ukraine). This doesn't mean that every crisis comes from these dudes.

If you have the capital to invest during a crisis, do it. I did the same, and don't feel guilty for it. More ethical to profit from the stock market rather than to make click-bait videos spreading disinformation about the pandemic.

I'm not going to tell you what I'm buying because I'm not a financial advisor. If you don't know what to buy, I suggest you hire a CFA, a certified financial advisor. You can even follow one on YouTube, just the way you follow the "alternative media"! My favorite one is Joseph Hogue of the channel Let's Talk Money![27]

However, I agree with you that if an economic system cannot cope with a one month stop, it must be restructured. However, no restructure is possible if people are not educated on how to manage their finances. Whatever system you apply, money will always end in the hands of who buys assets and will flee from the hands of who just buys liabilities. A simple example: Tom works as a software engineer. He's paid really well, something like $150,000 per year. He lives in a villa with a swimming pool,

---

[27]https://www.youtube.com/c/LetsTalkMoneywithJosephHogueCFA

4 bedrooms (one for him and his wife, two for their kids and one for the guests), a big garden and garage. He drives a Mercedes Class E and he always gets the latest iPhone. It sounds like life many of you dream of! However, if we look under the hood, Tom's financial situation is not a bed of roses. From his $150,000 a year, a good 40% goes away in taxes. Another 30% goes for his kids. Another 10% for fine dining and shopping with his wife. There's another good 10% which goes for the mortgage. The remaining 10% goes into material things, which Tom hopes will relieve him from the stress of family life. Nothing is saved. Nothing is invested. And let's ignore the possible loans to pay back! Tom owns a lot of non-essential things, liabilities, while he owns no assets, i.e. goods which produce money automatically for him. Despite his respectable salary, Tom is technically poor, because if he quitted his job for just one month, he would go bankrupt in no time.

Now let's see Jessica. She's a graphic designer. Jessica had chosen to work as a digital nomad, to save money. She gets something like $70,000 per year, less than half what Tom gets. Yet, Jessica is much wealthier than Tom. How? For 3 years, Jessica has been self-employed and moved to Portugal. Here, life is 3 times cheaper than in the US. If in the US, Jessica had to be careful at eating out, in Portugal she can eat out whenever she wants. Since Jessica is self-employed, she pays taxes only after she has spent her money. In Portugal she can even enjoy a 20% flat tax rate made by President Corona to attract digital nomads. But she won't pay that 20% over her $70,000. She will pay the flat tax only on what she didn't spend. With her lifestyle, Jessica spends around $40,000 a year, for her and her family. Jessica is going to pay 20% on 30,000 which is $6,000. Jessica still has $24,000 of savings a year[28]. With this cash flow, Jessica has already bought 2 properties to put on Airbnb. From next year, these properties will generate an additional $15,000 net a year. Can Jessica not work for a

---

[28]This is a simplified example. Tax calculations are always complex. Please, always rely on a certified accountant!

whole month if she wants to? Of course! Even longer! Then Jessica is technically rich. She owns at least two assets, making money for her, even when she sleeps.

Yes, I learned these theories from Robert Kiyosaki. Yes, I really respect the guy. I really believe his book should be in every secondary school library! Kids should learn his techniques from an early age. No, I don't agree with what he says about the pandemic. Robert is an excellent communicator, but he knows nothing about biology and medicine. He should be more respectful when commenting on topics which go outside of his expertise. Don't take for granted that who's really good at something is mister-know-it-all.

The Remote Work Revolution

The President forced millions of managers around the world to open their eyes over a terrible truth: their employees can work from home! And their employees can be as productive as when they were at the office! This was a blast for those who are affected by control mania and had to accept "to not see" their workers directly.

Unfortunately for these managers and entrepreneurs, this is the future. Of course, we won't see entire companies transitioning 100% remote. Not every role can be performed adequately from home. Some roles need a physical presence at the office, face to face exchanges, team meetings where you can feel your peers. Even those creative discussions on the floor in front of the coffee machine can be beneficial from time to time. That's where sometimes the best ideas stem from.

Many executives feel that the best performance can be achieved when the employees are in close contact in a physical space. The battle between flexible and fixed work settings is older than COVID-19. Yahoo was one of the first big companies ever allowing its employees to work from home now and then. However, when the new CEO Marissa Mayer stepped in, she ordered all the remote employees to come back to the office as *"people are more productive when they're alone...they're more collaborative and innovative when they're together. Some of the best ideas come from pulling two different ideas together."*

This remote work ban happened in 2013. Yahoo was already dealing with problems from the future! I would even dare to say that in 2013, Marissa Mayer was right. Back then, communicating and collaborating remotely was much more difficult than today. To call people we had to rely on Skype, which has always been unreliable by nature. After 10 minutes into the call, the connection started lagging, leading to frustrating experiences. When Skype asked to rate the call quality, you wanted to smash the monitor! Today we can rely on more stable tools such as Zoom and Slack, which helped millions of businesses survive during the pandemic. We even didn't have tools designed specifically to bring remote teams together, such as Microsoft Mural or InVision. Bringing ideas together today is much easier and more feasible than it was in 2013. This is why forcing people to return to the office today is nonsense.

When reading comments on LinkedIn, I see fear among the conservatives. These are not only managers or business owners, but even normal employees. These folks fear that if a remote work model will stay in place even after COVID-19, everyone will be forced to work from home. This mindset really pisses me off. Just because you people are used to impose your beliefs on others, doesn't mean the others want to impose their beliefs on you! Do you want to come back to the office? Do it! Just don't expect that everybody else must follow suit to feed your ego. Are you the only one in your company wishing to go back to the office? Change company! You can't expect that your colleagues will bend to your will. You might be alone in your company, but not in the world. According to surveys conducted around the globe, there is a good 20% on average (according to the country and industry) missing the office so much that they crave to return there from Monday to Friday like in the good old days. Another good 60% wants a hybrid model, so still making the office pivotal. The remaining 20% doesn't want to see the office ever again and these people must be respected as long as they are productive and deliver the work they are paid for.[29]

Don't you trust them? That's your problem buddy. If you don't trust your employees, they will switch to someone who trusts them. You will lose precious talent because of your own insecurities. The biggest mistake employers can make is ignoring their employees' needs. Giving the chance to work from home is like giving a handicapped worker an office easily accessible via wheelchair. You would never deny your handicapped employee to have a ramp instead of stairs to reach the desk. Would you?

Understanding this concept is essential. On LinkedIn, I had to remark this concept over and over again. It will never be Office vs Home! Logic and latest trends suggest we will have a wider choice than we had before COVID-19. Who will want to work from home, will work from home. Who will want to work from the office, will work from the office. Who will enjoy both, will embrace a hybrid model. Who will not respect these choices, will be wiped out like the dinosaurs!

Say what you want, but for me, working from home has been the biggest breakthrough! For me the most amazing part of working at home is that you don't have to appear! Just the idea of having to sort my hair, look fresh and make sure my clothes are fine is stressful. I believe for women the stress is even bigger. Think about all the time and money spared on makeup and cosmetics! What if you can just stay in pajamas or even naked and start your work day like that? That's awesome! Home is the only place where you can be ugly and enjoy it!

Beside avoiding the stylish chores in the morning, there are other more important perks coming with working from home.

---

[29]https://www.weforum.org/agenda/2021/07/back-to-office-or-work-from-home-survey

The best impact in my case was a steady improvement in my daily habits. When I was at the office, my breaks were everything but healthy. As I don't smoke, I seldom joined my colleagues on the terrace, preferring the kitchen. But an office kitchen is full of unhealthy temptations: snacks, chips, chocolate bars, hot chocolate, cappuccino. Some days I was capable of drinking even 4 cappuccinos out of boredom, just to have something to do during the break, all while my belly was growing and my athletic shape was dropping...

Since day 1 at home, I replaced those harmful breaks with workout breaks! I started doing some push-ups and some abs crunches every time I took some little time off the screen. After a few weeks, I raised the bar and started to follow Chris Heria and Next Workout on YouTube, beginning to properly exercise from home. I wasn't alone. Millions of people adopted this lifestyle and started to integrate exercise in their work routine. Even in my company, we started to organize Zoom classes, including yoga and Pilates. With all the communication channels we have today, everything is possible and if you don't do it, then it's because you're finding excuses to not do it!

Physical shape aside, I also appreciated how working from home brought a new flexibility into my life that I didn't know before. If I need to go somewhere, I just go and I'm still able to recover my time from work the very evening of the same day. If I ever have any doubt about whether I've left something unfinished, I can quickly access my laptop without having to go back to the office.

This flexibility helps me manage my life more easily. The elimination of commuting added up a bonus of precious time. It's so sad how much time of our lives we waste commuting to our workplaces. Time that will never come back. I was already lucky that I used to walk to my office, but still it was a walk I didn't enjoy much. It took me like 20 minutes to get to the office and I had to pass through the hub of Malta's nightlife, Paceville. Going through it in the morning meant to witness the staff cleaning the rubbish left from the night before. Going through it in the evening meant slaloming into the crowd that was gathering for another crazy night, let alone the annoying hookers who tried every time to drag me into their gentlemen's clubs.

But the part I miss the least is the area around the office, Spinola Bay. It is noisy, heavily polluted and the healthiest bite I could grab from there was the crunchy tuna ftira from Tony's Bar. The tuna ftira was recently declared UNESCO Food Heritage and I agree with it! A special crunchy bread stuffed with nothing but onions, tuna, olives, tomato paste and dressed with pure extra-virgin olive oil and black pepper. It's healthy of course, but I couldn't eat that every day. The basis of a healthy diet is variety. This variety used to include fancy delicacies like sushi or Indian food, for the joy of my wallet, otherwise the alternative was any kind of pizza baked by the nearby restaurant, Marobbio. In this place pizza is amazing, but when you want a six pack, eating this dish even 3 times a week doesn't help reach the goal. Thanks to President Corona, I reduced my pizza intake to a more reasonable rate of once a month. Now like magic, the six pack is there....

Talking about food, the new remote work configuration was the chance for cooking again. I love cooking, but business life took me away from the kitchen most of the time. Finally, I could cook my own food again and choose my own ingredients. A good physical form is more achievable if you eat clean, simple food. A good plate of lentils dressed with nothing but extra virgin oil, served with a boiled egg, gives you all the proteins you need for your next workout. A clean salmon steak dressed with a touch of rosemary and balsamic vinegar is as tasty as a kebab stuffed with who knows what kind of sauce... Ah, and don't forget vegetables! Possibly raw! Oh, I love raw vegetables! At first they may result tasteless or even weird, but with time your mouth will appreciate that sense of freshness! And you know what's the tastiest raw vegetable?

Artichokes! These spiky green gems are my favorite thing about winter! You peel the outer layers, then you start eating the white part of the inner "nails" until you reach the artichoke's heart which is so crunchy, delicate and tasty! Eating this flower (it's a flower indeed!) is a little journey!

Of course, in my kitchen there is also room for fancier stuff. The feat I'm most proud of during 2020 was when I resumed one of my biggest specialties: lasagna. But I didn't simply bake lasagna again after years, I reinvented this dish in an animal friendly fashion! After some feasibility assessment, I found out that making a real Bolognese sauce using Beyond Meat was possible! Beyond Meat is a patty made out of peas protein, potato starch, beans and beetroot. They created the first vegan burgers that taste exactly like meat and that do not employ soy. Beyond Burgers had already been part of my diet for some time, so it didn't take long for me to recycle their patties in my Bolognese sauce, replacing dead animals in it. Seeing my friends' faces after telling them that my lasagna didn't contain meat was priceless. I still remember the text my cousin Michele sent me after he saw my creation on Facebook:

"When I saw the vegan lasagna I wondered: WHY??? But then I saw that many pretty damsels appreciated the result and then I thought: WELL DONE!" :-D

Whether you like it or not, remote work is here to stay. The President's decree is already definite law. Nevertheless, there are ample pockets of resistance. I witnessed businesses which illegally forced their employees to keep going to the office. Beside these dirty examples, you have control freaks who panic if they don't see their employees near them. Visual contact is vital for them. I suggest these employers to see a good therapist if they don't want to lose their best talent for a mania which is easily curable.

Another faction within the office resistance is formed by the very employees though. It's no secret some employees hate to work from home. Working from home is not the ideal for everyone. The most extroverted employees might need the coffee and/or the smoke break with their peers. They might feel the need to engage more in live meetings. Their roles might require a physical presence at the office. Others simply get tired, lazy, bored or even annoyed at home. It's normal they feel threatened. That's why we must make sure that the remote work revolution doesn't become a radical revolution. Guaranteeing the freedom of choice is a priority. Companies should always be able to arrange a minimum of office space for who needs it. The right to work from the office is as important as the right to work from home.

The other major pocket of resistance comes from who you wouldn't expect. In Silicon Valley, some giants expect their employees to come back to the office. Google is the most prominent firm to pursue this dream. While acknowledging the importance of a hybrid model, Google CEO Sundar Pichai made clear that the company expects all the Gs to live nearby the office. Forget about living in fancy tropical islands while managing Google business. For the first time, Silicon Valley appears conservative. Or maybe it reveals what the tech industry is about: selling apparent change while defying real change.

The cost of working from home

Like every revolution, even the remote work one demanded its toll. The bars and restaurants who populated our city centers bankrupted. Those who survived are panicking now. They want workers back to the office so they can spend the lunch break at their tables, tasting their junk processed food, predictably at 1.00 pm every day from Monday to Friday. The landlords from the City of London even lobbied (or pressured) the British government to force people back into the office to preserve the value of their properties. Having an office in the City must keep being crucial. Renting a residence in the City must keep being a luxury that people are desperate to pay for, giving more than 1/3 of their hard-earned income to lazy billionaires.[30]

Landlords of New York even shouted that going back to the office is a Patriot Duty! If people don't return to work, many bars and restaurants will die. Sure they will, but other bars and restaurants close to the employees' houses will grow. Money never disappears, it moves.[31]

Landlords and property firms displayed arrogance and insensitivity. With the pandemic far from over, they have been pushing authorities to speed up the return to the office, thinking only about themselves. A behavior that you cannot certainly define, patriotic...

Remember, you owe nothing to landlords, bars and restaurants of your city. If you feel better working from home, do it! Fuck them off! They don't own your life, even if they love to think so.[32] All the money we saved during the pandemic cannot go into the pockets of the same few arrogant jerks who play Monopoly with our cities. Let's use this surplus to boost the economy with more important things than a shitty, overpriced studio apartment in Shoreditch! Let's spend this money on assets, on valuable professional courses, for the gym or for a compelling, epic trip. This will help us appreciate freedom more than ever and find our real selves again!

---

[30] https://www.standard.co.uk/business/property/central-london-office-shops-wfh-b920900.html

[31] https://nyctalk.org/new-york-city-landlords-ask-bosses-to-speed-up-the-return-to-work-process/

[32] https://www.bloomberg.com/opinion/articles/2020-12-01/get-ready-for-a-supercharged-economy

End of Capitalism?

In the early stages of the pandemic, we heard and read all the most diverse apocalyptic scenarios: "This is the end of the world - The Bible foresaw this - The Antichrist is coming - We are going back to the Middle Ages!", and so on... But the most prominent idea, even among tough intellectuals, was that capitalism was going to die, guilty of being an unfair, short-sighted system incapable of dealing with daunting problems like climate change or a sudden pandemic. President Corona was going to declare capitalism a failure. On social media, we saw a strong resurgence of left-wing fans, excited by the prospect that the economy will be finally planned by the state! By contrast, this fueled fear among others, who began to speculate that the pandemic was just an excuse to create the perfect situation to reestablish communism once for all. These speculations were well received by authors like David Icke, who amplified the idea that a red plot was in action, all around the world.

Capitalism has always been divisive. People are never sure how to relate to it. Many hate it, few love it, most tolerate it. Even if it ends in "ism", capitalism is not a precise ideology, like communism or fascism. This blur makes it harder to define it. According to Investopedia:

"Capitalism is an economic system in which private individuals or businesses own capital goods. The production of goods and services is based on supply and demand in the general market — known as a market economy — rather than through central planning — known as a planned economy or command economy.

The purest form of capitalism is free market or laissez-faire capitalism. Here, private individuals are unrestrained. They may determine where to invest, what to produce or sell, and at which prices to exchange goods and services. The laissez-

faire marketplace operates without checks or controls.

Today, most countries practice a mixed capitalist system that includes some degree of government regulation of business and ownership of select industries."

We always hear that capitalism is making the wealthy wealthier and the poor poorer. But, is it really like that? According to surprising studies, this statement is very far from the truth.[33] If we take the $1.00 a day poverty threshold, we can see that the world population living below this income has basically disappeared over the last 180 years. If we raise the threshold at $1.90 a day, we can see that, at the rise of the industrial revolution, more than 90% of the population was below this wage. At the beginning of the 21st century, even this kind of extreme poverty vanished. There may still be a good 50-40% of the world population living with just $2.00 a day, but it is still in sharp decline.

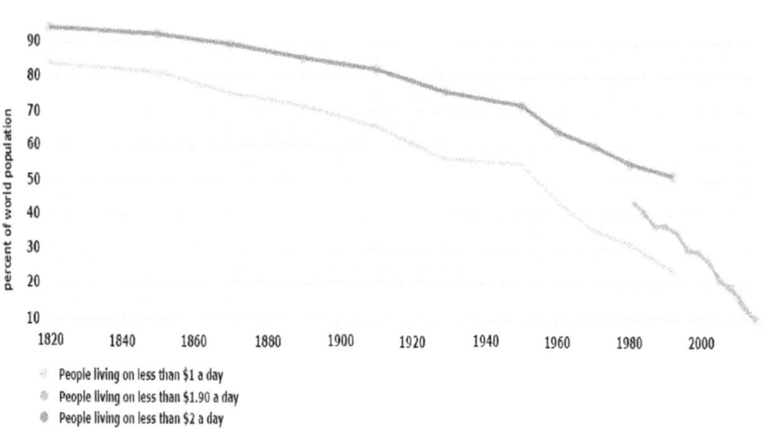

[33]https://www.humanprogress.org/five-graphs-that-will-change-your-mind-about-poverty/

OK, this graph might shock you, but if you think carefully about history, it all makes sense. Let's go back to the Middle Ages. Even the richest medieval king would sell his soul to live in your modern apartment. Nowhere in the Middle Ages you could find such a place which is so warm during winter and yet so fresh during summer. The bed, so soft, yet stiff enough to support your spine. The kitchen provided with all the tools to make the work of 100 servants. The living room is more functional than any throne room, equipped with the most powerful of the magic tools, a special crystal that allows you to see what happens afar! Then another powerful crystal which encapsulates as much knowledge as the Dresden Library! Last, but not least, a horn which helps you get in touch with all of your soldiers, at any moment!

We don't even have to go back so far. You can just remember what your grandparents, or your great-grandparents if you're lucky enough, used to tell you about their youth. Even in the West, many grandparents had a childhood spent in poverty. Most of them used to wear the same clothes and the same shoes every day. Showering was not an option. They bathed with boiled water to throw on themselves, shivering once wet if the bath was in winter. If they needed the toilet, it might have been a hole in the ground some yards away from their house. To clean their butts, the best option was the newspaper from the day before, not the softest material. No surprise that hemorrhoids were a frequent curse in the old days.

Traveling was out of discussion for the poor. If traveling ever occurred, it meant that it was for good, bound to a richer land where to find fortune. It was a real goodbye, even because most of the traffic was still by boat and we all know that the ocean is not as friendly as the sky... If your big brother left for another land, letters were the only viable way to keep in touch, provided that you didn't receive the cruise line's letter announcing the sinking of your bro's ship...

Career paths were limited and you had to choose from a tender age what you wanted to in life, if you could choose at all. Studying wasn't for everybody. It was expensive, even in Europe, where nowadays, study is a fundamental right and education is free. If you could choose, then you couldn't change, not easily at least. A doctor remained a doctor for life. A lawyer, always bound to the same court. A baker always baking the same bread. Our ancestors' menu wasn't the most varied.

A radical shift was possible thanks to the unprecedented wealth created by capitalism. If we put this concept on a graph, you will agree right away:

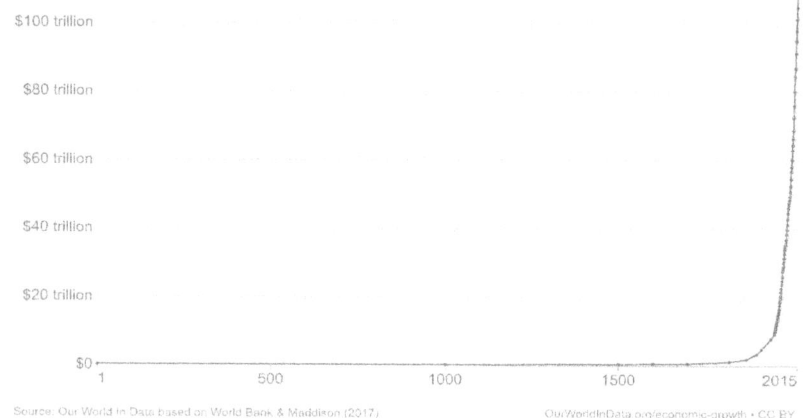

Look at the increase in the global GDP from 1800 till 2000!!! In 200 years, capitalism created more wealth than in all the previous human history!

The same outstanding increase applies to scientific and technological progress too. Before 1800, the scientific and technological progress had a low, stable pace, with few significant breakthroughs every century and were usually brought up by rare gifted minds. When capitalism made its way within society, innovation sped at an exponential pace. Most of the tools that we use today, would have never come to life without the innovative force that capitalism brings. When you have entire companies competing for technological supremacy and big teams working on a project, rather than one single Leonardo da Vinci, innovation is accomplished more easily and way more often. Many capable individuals working together are more powerful than one genius individual.

How is it possible that today so many people are capable of working on science, when till a couple of centuries ago, science was stuff for a very restricted elite? Well, capitalism helped knowledge to spread faster than any public program. Thanks to TV first, and the internet then, scientific knowledge became public domain. Nowadays, even the poorest individual can access courses from the most prestigious universities in the world. Platforms like Coursera are one of the most beautiful examples of this trend. All you need is an internet connection, that again, more and more people are capable of affording as time goes by. Imagine you could tell a medieval farmer kid: "Hey, what if I gave you a crystal ball that would let you know what only the wizards are allowed to know?". It sounds like those advertisements which say "Watch this video before THEY take it down!" or "What THEY don't want you to know!". Today this concept is used for marketing purposes, but before capitalism, ignorance was the natural condition for 99% of the population. It was an effective way for the rulers to keep their people on check. Not that it was intended. After all, even the kings knew little themselves. Usually kings relied on the "knowledge" from the priests, a batch of information often coming from a "higher power" through "divine" enlightenment.

Enlightenment is the other fundamental word that made capitalism the innovation force it is today. In 18th century Paris, something magical happened. An elite of high intellectuals stopped relying on divine, unproved knowledge and began to investigate and question the reality around them. This is the movement and the age of very cool folks like Emmanuel Kant, Adam Smith, Cesare Beccaria, Montesquieu or Voltaire! Thanks to brains like these, today we enjoy the clean separation between Church and State, we appeal to a law which works the same for everybody (on principle at least!), we are encouraged to question what happens around us, we live free from any sadistic absolute king and his brutal knights, and we run a free market economy where the most important asset is our merit and hard work, and not the family name or the social status we start our lives from (still on principle, at least).

Could capitalism have ever existed without the Age of Enlightenment? There's much speculation about this question, but most agree that it couldn't. Not only capitalism, but any other modern ideology like socialism and communism and the modern world itself. There was however, one particular institution that was already working like a modern multinational enterprise, long before capitalism appeared. The biggest company of all time, capable of enduring over 2 millennia and conquering the market in all the continents. I'm talking about the Catholic Church, the very institution that was harshly criticized and even weakened, by the intellectuals of the Age of Enlightenment. Before the modern world, the Catholic Church was the only company effectively moving capitals around, storing them, accruing them and reinvesting them into something else. Even the biggest public works, the construction of churches, were the most common events that generated jobs and propelled the economy. Everywhere, there was just "one ecumenical holding company..."[34]

Yes, we would still live in a world dominated by those priests or shamans who knew very little. Today there are still people and communities like that. I saw a lot of comments against COVID-19 vaccines coming from religious fanatics, wishing the doors of hell for the vaccine makers. Even today, there are still folks relying on prophets, like Jake Angeli for example. After the US Capitol Attacks, Jake claimed to be an alien, bound to ascend to a superior reality. He sees truths that the others cannot comprehend.

---

[34]https://www.youtube.com/watch?v=yuBe93FMiJc

It takes a lot to change, for everyone and everything. Even capitalism itself evolved so much, since it appeared as we know it, as the socio-economic system built around the industrial revolution. If you think for a second, 19th century capitalism has nothing to do with today's capitalism.

In the 19th century, the relationship between the employee and the employer was something closer to slavery. Today, employees and employers base their relationship on mutual respect, leading to a real exchange of services.

In the 19th century, employees had little rights and no alternatives. Today, employees have all the rights and many alternatives, if they don't like their job anymore.

In the 19th century, employees used to work 12, even 14 hours a day, doing a repetitive job under harsh conditions. Today, employees work 8 hours a day, 1-2 hours of which are dedicated to lunch break, cigarette break, PlayStation break, flirting break and so on. In some countries, they are even experimenting with the 6 hours work-day! This would have sounded like a Jules Verne novel if you told a 19th century worker!

In the 19th century, people begged to work in a dirty factory. Today, businesses do everything to make their work environment more alluring to potential employees. The effort is flipping sides.

Of course, a business of the 19th century had only one, ruthless and super-rich owner, only profit-oriented. Today, employees can own shares of the company they work for and if they're on time to board the business in its early stages, they may even become millionaires out of their own work. After all, this system was the ultimate goal of the socialist unions at the beginning of the 20th century: letting employees own the business they work for. This socialist dream became reality thanks to capitalism 3.0. What a beautiful paradox!

Thinkers like Russell Brand, state that capitalism makes everyone poorer and poorer and the very few richer and richer:

"Our system, capitalism, is designed to behave like this: It generated wealth for the wealthy and further impoverishes those with nothing. Asking it to behave differently is like asking a microwave to wash your car."

Brand started from a simple conclusion. Capitalism would be based only on profit, and "wherever there's a profit, there's a debit". This is partly true. If I buy Brand's book *Revolution*, I'll get £8 poorer and he'll get £8 richer. That's what it looks like if we take this transaction alone, without the bigger picture that an economic system is. Thanks to my money, Russell will be able to buy other goods, like fashion clothes, flight tickets or drugs. This will make other people richer and with that money, they will buy new services from others. At the end of this long chain, someone will buy my book in turn and I'll recover my £8. The faster this process happens, the more people enjoy money. Wealth is not about how much capital you have, but about your cash flow. The cashflow is the amount of income in a unit of time (day, week, month or year). For example, if you have a cash flow of $10,000 a month ($120,000 a year), that's a really good cash flow. If you win $100,000 through a lottery, that's not so good. You get $100,000, but only once in your lifetime and for sure you will burn that capital at a drop of a hat. The flow must be constant to label you as a rich person.

Imagine you have 6 kids playing with a ball. Every kid wants the ball, but every kid also wants to pass the ball to keep the game alive. The faster the ball moves, the more the kids will be able to enjoy the ball (economy speed). This means that the kids' cash flow will be higher, thus they will be happier (richer). To reach the perfect economic balance, the kids should make sure to pass the ball evenly across all the 6 members (wealth distribution), so that every kid is happy and doesn't feel excluded (poor). This principle propels one of the most debated aspects of capitalism: consumerism. To keep the economy alive, economists push people to buy more and more, even and mostly, things they don't need. The whole society is built around this concept. We are literally bombarded by advertisements, every minute of our day. Companies are investing billions in algorithms capable of predicting what we want to buy. If we buy nothing, the economy dies and everyone becomes poor.

If the world was just a group of kids, consumerism would be harmless and would make sense. Unfortunately, the world is more complex and delicate. Let's add to our kids' game that every time a kid touches the ball, a tree around them burns. If they pass the ball too fast, all the trees around them will burn, leaving the kids without protection from the excruciating sun. Because of the unbearable heat, the kids will soon be forced to interrupt the game, altogether. This is the main problem with consumerism. It requires us to consume all the resources available, as fast as possible. If you think for a moment, it is obvious this model is not sustainable in the long term. It's not about climate change. Even if you don't believe in climate change, the problem of limited resources remains. Economist Nicholas Georgescu-Roegen already proposed in 1971 that the economy cannot escape the laws of physics. Being a species confined on a planet which can offer only a limited amount of resources, we cannot dream of consuming them indefinitely, with an ever-increasing pace. It's obvious that even the concept of infinite growth is not sustainable, not in the physical world. A GDP based on natural resources exploitation will not last forever. It will reach a breaking point sooner or later. This has already happened in human history. You can find a shocking example by reading what happened to the people of Easter Island, when they put down all the trees in their land, to build their iconic statues. The statues survived their creators...

In Russel Brand's vision, the only solution to this madness is a socialist system. However, as history has shown on multiple occasions, that would be a horrible solution. A more reasonable (and bloodless) solution could be expanding our economy into digital worlds. This is already happening. Virtual

Estate is already a thing, bound to increase exponentially over the coming years, thanks to the deployment of the blockchain technology, the oracle capable of creating unique digital entities, the so-called Non-fungible Tokens (NFTs). NFTs allow you to sell a virtual villa in a virtual world for the same amount you would get from its bricks and mortar version. All without the harsh environmental impact required by the construction of a house. It's fair to call it Virtual Real Estate. The goods are virtual, but the profits generated by them are very real!

If you're an unemployed architect, you may want to look at the virtual worlds that are being built on the blockchain. *Decentraland* is a perfect example, where you can create and sell virtual estates. Another intriguing alternative is called *Somnium Space*, a VR metaverse, fully immersive. It sounds a bit scary, like the Matrix, but these worlds can create massive opportunities. They can expand the real estate industry, without the need to impact the environment. Virtual worlds can be the greenest thing ever conceived!

Real estate is not the only good tradable in a metaverse. Already since the times of *Second Life*, people can get jobs in these virtual worlds. Second Life entrepreneurs used to pay their employees with Linden Dollars, a digital currency created before Bitcoin. Today, the blockchain allows people to get paid in juicy cryptos or even real Dollars and Euros. In these metaverses, the only limit to growth will be people's creativity! Once again, the future seems to be digital, where the Real World and the Matrix dance together...

Humans had always sought profit, long before capitalism appeared. The search for profit embedded into capitalism is not so bad per se. Even Mr Brand profited from the sales of his book and used these profits to do something very good. He opened a cafe which employs recovering drug addicts. This great initiative gives these people the chance to feel part of society rather than criminals. This is wonderful and was possible to profit! Russell shouldn't be ashamed of this, but proud. Profit is what moves society, regardless of the system adopted. If society hasn't collapsed yet is because most of the businesses out there seek legit, ethical and honest profit. Did you know that in the US, 1 out of 7 companies commits fraud? That's a lot, but this also implies that 6 out of 7 companies do not commit fraud and carry their business with honesty and integrity. You're not the only one who cares about integrity, arrogant dummy...[35]

We saw that capitalism drags entire populations out of poverty, boosts technological progress and makes goods cheaper and more accessible. However, it tends to consume the very planet we live in and if uncontrolled, it can lead to the planet's destruction. With a bit of common sense, we can agree that capitalism needs further improvements, it needs proper, balanced regulations and in the future, possibly, society will move past it, maybe to something else we can't even think of today.

For sure, history showed that communism is not the alternative. It makes people equal only on paper, while stripping their freedom in practice.

I've always wondered why there was never a free democratic communist country but there were several free democratic capitalist countries (as free as a complex society can be). I personally believe that the answer is much simpler than you might think. Yes, this time around we don't need to go through lengths to explain why communism is incompatible with freedom and democracy in particular.

Sir Richard Branson once said: "Everyone is born an entrepreneur!".

This quote says a lot. We all, as human beings need our personal space, we all need a personal dream to achieve and we need to manage our business the way that matches the most with our values, habits and beliefs. Here business is meant as any activity that produces some kind of "profit" for

---

[35]https://www.youtube.com/watch?v=MV9KzOS9MGI

us, not necessarily in the form of money. We need profit as much as we need private property. This feature is so natural that even animals like cats and dogs need to OWN their toys and their territory.

Even the People's Republic of China had to welcome private property and a free-market economy to survive. China is still communism in the institutions, but the economy is more capitalistic than ever, after the reforms occurred in the 1980s, by the will of then paramount leader Deng Xiaoping. Since China completed its shift from a planned economy to a free market system, the Dragon became that unstoppable productive force capable of producing most of the items we use daily. The Made in China brand became ever-present in every corner of the world. And look at China's poverty rate sharp decline from 1990:

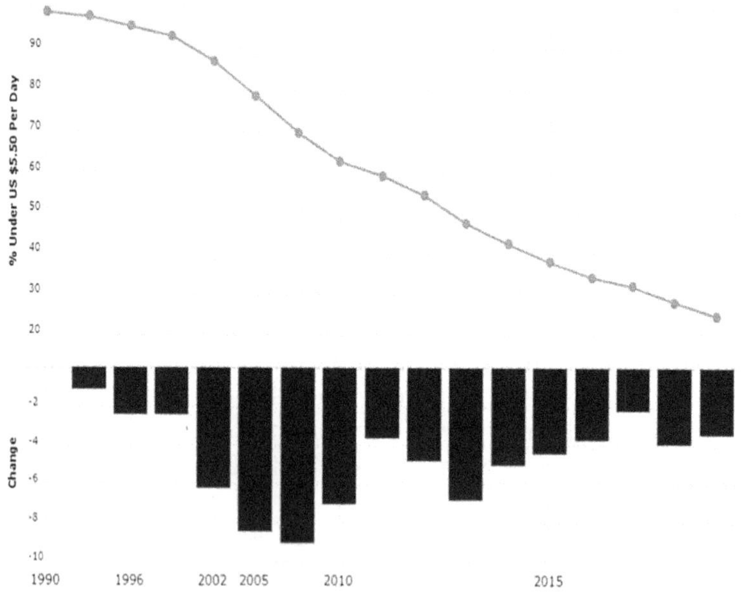

China Poverty Rate 1990-2023 [36]

[36] https://www.macrotrends.net/countries/CHN/china/poverty-rate

This confirms once again how capitalism is so effective in reducing poverty! By understanding these factors, we can conclude that President Corona won't kill capitalism. All the other way around! He made capitalism even stronger!

Capitalism must tackle two major problems to make a serious leap forward, though:

- Ultra-billionaires
- Financial education

The ultra-billionaires problem

Ultra-billionaires are those individuals like Elon Musk or Jeff Bezos who possess a wealth comparable to those of entire nations. This fact alone poses a serious question. Is it right that an individual is so rich beyond any limit? Could this person turn into a James Bond villain? That's what Musk and Bezos look like. They remind me of Hugo Drax, the main antagonist in *007 – Moonraker*. Drax is a billionaire said to have bought even the Eiffel Tower. When Bond flies over Drax's estate, the pilot says:

"Everything you see belongs to Mr Drax. - He owns a lot. - What he doesn't own, he doesn't want."

Drax's main business is manufacturing space shuttles for NASA (what a creepy coincidence!). His desire to conquer space hides a dark master-plan: to wipe out the entire humanity and repopulate the planet with a group of humans personally selected by Drax himself, who match his ideals of beauty and perfection. In his plans, these chosen ones will spend some time with him on his secret space station, waiting for a special toxin to kill every human on Earth (sparing any other form of life). The survivors then, will return to Earth to start a new, perfect race. Drax wants to decide the destiny of humanity, on his own. This arrogant trait matches most tech CEOs. Their aim is to reshape the way humans live and their future, even when we don't want to live their way. Because of these CEOs, nowadays we are constantly tracked, spied on, invaded. We are surrounded by technology which gives us orders and controls us, when it should be the other way around. They decided that you can't live without your phone and that you can't access any service without it. They decided that in the near future, machines will do everything humans can do, but better, stripping us from the joy of doing something. They even have already decided that your grandchildren will merge with those machines! Elon Musk's motto is indeed: "If you can't beat'em, join'em!"

Yeah sure, they will defend themselves saying that nobody is forcing you to adopt their technology, but seriously speaking, what kind of life can you lead without a smartphone today? Without a smartphone, you can't work, you can't pay, you can't access any service, you can't even date! You're out of society!

All of these important decisions are constantly made without any public consultation or any democratic vote, of course! Traditionally, ultra-billionaires don't like democracy. They believe themselves to be superior beings, so their judgment is far better than the average voter. Because of this, they should be able to skip the line and influence governments directly. In Russia we saw this trend with the oligarchs, who took over the country right after the USSR's collapse. Over the years, Putin tried to get rid of them, with partial success. Most of them still control the major Russian industries, keeping all the wealth for themselves and leaving breadcrumbs for the other Russians.

In America, ultra-billionaires set up an ingenious system, since the nation became an industrial power: **Lobbying**. Lobbying is genius because it's nothing else than legalized corruption. A big corporation may offer to fund a party's campaign and in return, if elected, the party will do what such corporation demands, like cutting taxes for its sector, loosening regulations or changing them to their favor. The system became so ubiquitous over the decades that nowadays, both the Democratic and Republican parties are often backed by the same companies! Most of them belong to the golden list of Fortune 500, the most profitable and hence, influential companies in America.

If we look at just a few examples, we can notice how, over a 10 years period (2007-2017), the amount donated to the two parties is almost equal.[37]

Few examples:

Lockheed Martin

Sector: Aerospace and defense company Lockheed Martin is based in Bethesda, Maryland.

Total donations: $2,364,432

Percentage donated to Republicans: 54.7%

Percentage donated to Democrats: 43.7%

General Electric

Sector: Heavy industry, generators, turbines, aerospace engines, mining equipment.

Total donations: $1,753,871

Percentage donated to Republicans: 43.2%

Percentage donated to Democrats: 56.0%

Delta Airlines

Sector: Airlines. Delta is one of the major airlines in the US.

Total donations: $2,045,942

Percentage donated to Republicans: 52.2%

Percentage donated to Democrats: 45.4%

[37] https://www.businessinsider.com/fortune-500-companies-republican-democrat-political-donations-2018-2#exxon-mobil-8

Exxon Mobil

Sector: Oil extraction, refinement and distribution.
Total donations: $2,134,633
Percentage donated to Republicans: 63.0%
Percentage donated to Democrats: 36.0%

What about Big Pharma? They are generous donors too! Don't worry!

Pfizer

Total donations: $1,943,348
Percentage donated to Republicans: 34.8%
Percentage donated to Democrats: 64.3%.

Johnson & Johnson

Total donations: $1,981,003
Percentage donated to Republicans: 45.4%
Percentage donated to Democrats: 52.2%

What we can notice is the striking balance of the donations over time. Why? These companies are not really interested in a party's ideology. They simply want to secure the loyalty of who's going to win.

"If you win, you won thanks to our financial support, so don't forget about our needs and mostly, don't tax us, buddy! We will be your priority, not the Americans who voted for you..."

Congratulations, dummy! Even this year, you voted against your interest!

Yes, there are also unbalanced instances, where a company pledges undying loyalty to one party. This comes also from tradition. Oil companies have always been closer to the Republican world. The Bush family has been into the oil business for decades, after all. Tech companies on the other hand, tend to prefer the Democrats and by far. Apple pledged

87.5% of its donations to the donkey, while Facebook's contributions topped 91.3%. That's probably why Trump was so active on Twitter...

When conservatives state that the tech and the entertainment industries are biased, they are not completely wrong... However, the conservatives' whining is still pointless. The problem is not about being conservative or liberal. The problem is that both the Democratic and the Republican parties pursue the interests of giant groups and ultra-billionaires, who are so distant from the average American. They will never put effort in solving your daily problems, because they can't even understand them and because, they have "higher" causes to pursue. What do you expect, when your parliament is composed of people, whose poorest has a 6 figures bank account?

Why are rich congressmen a problem? The problem is not just them. It's about the culture and fairytales that wealthy business people create around capitalism, around the concept of the American Dream. They tend to dupe you that everyone can be like them, with enough hard work. Elon Musk always tells how he was bullied at school, but thanks to his curiosity, passion, and hard work, he overcame all those obstacles and became a billionaire. Jeff Bezos loves to remind how he started from nothing, how brave he was to quit his corporate job and risk everything into his creation, when Amazon was just a server in a garage. They would be the very embodiment of the American Dream. If you work hard enough, there is no limit to what you can achieve! Too bad this view is a distortion of the American Dream. This ethos was postulated in the Declaration of Independence, where it states that *"...all men are created equal, that they are endowed by their Creator with certain unalienable Rights, that among these are Life, Liberty and the pursuit of Happiness."*

This statement means humans have the right to live a fulfilling life, independently from their social class, and most importantly, regardless of where they are born. This last concept is fundamental. It doesn't matter if your parents are poor, toxic, criminals or whatever. You should not be judged based on your heritage, and no barrier should be put between you and happiness, the same happiness experienced by children born in normal families. It's a concept which today might sound obvious, but that wasn't in the 18th century. Back then, your family influenced all of your life. If you were born in a family of peasants, you were expected to be a peasant too, for the rest of your days. Whoever dared to change this status, was met by strenuous resistance. Social mobility was an exception, not the rule. Social mobility is still a problem in the modern day world, in any country. The situation has improved dramatically, and social mobility is not an exception anymore, but there are still huge obstacles for who is born in the wrong place. If you're born in the US, you already have a huge advantage compared to who is born in Afghanistan. However, who is born in the US, still doesn't have the same chances everywhere, especially who's born in the "wrong" place or even with the "wrong" skin color. Who's born in a poor family, will see access to good education barred most of the time. Oh yes, you can win a paid scholarship, but you must prove to be a genius, while rich kids will have access to good education even when they are dumb. This system of private education for the rich is so rooted in America, because the country is run by billionaires and for them this is normal. That's how they can't conceive that in Europe education is mostly free. They don't even try to understand it, ruling the European system as socialist, with superficial disdain or even mockery, like it happened on Fox News. On the popular "news" channel, Trish Regan tried to compare Denmark to Venezuela. According to this "shrewd" host, Denmark's free educational system would have "stripped people of their opportunity" and that "nobody wants to work! This is the real problem!", too bad that Denmark scores 11 points higher than the US when it comes to employment...[38] [39]

What was appalling about Trish Regan weren't the falsehoods about Denmark, but how free education for everyone would be a horrible thing! How can she view it that way? It can be bad only if you have an aristocratic vision of society. Because this American elite is not even capitalistic, but is resembling more and more an aristocracy, where money and power must remain within the same few, possibly even within the same bloodline. This mindset betrays the very Declaration of Independence.

Lobbying does the rest. It ensures that the same kind of political forces are in power, without any disruptive party playing a real opposition. If you take the magnifying glass, Democrats and Republicans are quite the same, if not even two faces of the same party. Just look at the PRISM program, the huge mass surveillance system uncovered by whistleblower Edward Snowden. PRISM was started by President George Bush (Republican) and ultimated by President Barack Obama (Democrat)...

Both Reps and Dems hunted down Julian Assange, the man who dared to uncover what the West secret services actually do. And how to forget when both parties voted to invade Iraq in 2003, despite the total lack of evidence of bioweapons in the hands of Saddam Hussein?[40]

PRISM was just the last of a series of policies to turn the US into a police state. Both parties deployed an increasing number of police officers in the streets. Incarceration rates increased steadily under any American government, reaching a staggering 639 inmates per 100,000 people in 2021, the highest in the world, higher than third world nations.[41]

---

[38] https://www.youtube.com/watch?v=JXecLXlzEXE

[39] https://data.oecd.org/emp/employment-rate.htm

[40] https://eu.usatoday.com/story/news/politics/2019/04/11/julian-assange-six-wikileaks-most-memorable-revelations/3434371002/

[41] https://worldpopulationreview.com/country-rankings/incarceration-rates-by-country

The disastrous war on drugs was carried with fierce drive by both parties, treating addicts like inhumane criminals rather than patients. And both parties, of course, brought military spending to astronomical figures (we will cover this very topic in great detail later on...).

See? Whoever you vote for, things will still go in the same direction. It will always be a powerful bunch of lords busy to keep power and scared by their own people, so scared they feel the need to spy on them 24/7 (PRISM).

Differences between Democrats and Republicans are minimal, usually regarding aspects like aesthetics and cultural or moral beliefs. Beside those, you can just look at the political debates between their leaders and you will understand that the political content is null. Media focus more on how candidates are dressed than what they have to say or what these persons have achieved in their lives. The media and political systems, which endorse each other, are fine with that, as long as there is no actual third party to disrupt... the party![42]

Have you ever watched Alien vs Predator? This movie has a powerful tagline that should make you think: "*Whoever wins, we lose!*". That's exactly what happens every 4 years in the US. People cast their vote, choosing either the Republican Party or the Democrat Party.

I like to think of Republicans as the Aliens. They are more instinctive, ready to attack whoever threatens them, storm the opponent. Like the Xenomorphs, they have a strict hierarchy and they stick to precise values and behaviors, often inducted by their strong religious beliefs.

Democrats look more like the Predators. They are still ready for violence, always searching for new planets (nations) to invade, but they are also sophisticated. They never take an action out of the blue, but come with a plan after long discussions where all the high rank predators are involved. Their sophistication can be noticed even in how they win votes by exploiting the star-system, with Hollywood and music stars

---

[42]https://www.youtube.com/watch?v=6LPuKVG1teQ&t=1s

always ready to work for them.

So can we conclude that both the Democratic and the Republican parties nowadays are two big mountains of rubbish? Yes! So why do you keep on voting for them??? Are you retard? Yes! So once again, we can say that your worst enemy is just yourself. Once again you are the problem! But even the solution!

That's why you should start doing the unimaginable at the next elections. Remember, political parties are just a tool. When a tool is outdated, buy a new one... Yes! I dare you to vote for something you've never voted for before! I even dare you more! Why don't you guys try to create a new political party? In Europe, this is very common and indeed, parties change over the decades. It's rare to have a 200 years old party still running in a 21st century election. It shouldn't be the case, because society evolves, as well as people's needs, fears, expectations and hopes. America went so close to create something new with the Tea Party in 2009 and the Occupy Wall Street movement in 2011. Try again. Maybe Kanye West might lend you a hand. I'm serious. The last presidential election debates showed that Mr West cannot be any worse than a Trump or a Biden. I feel like advising Kanye West on how to fund his presidential campaign for 2024. He might start a special ICO (Initial Coin Offering) on the blockchain, to fund the campaign and attract all the crypto-wealth out there. If West does that, he will grant funds and votes of almost every crypto-enthusiast and with that money, combined with his popularity, the votes from the other Americans will pour like the Niagara Falls!

Financial education

However, a new political party's shaking effect on society will be short-lived, if society doesn't learn how to use its money. The other big root cause with money conflating into the 1% of the population comes from the fact that this 1% knows how to make money work, while the other 99% doesn't.

The poor work for money, while the rich make money work for them.

99% believe that the only way to make money is going to work, do whatever their boss tells them and hope to receive a salary every month. They will use that money to buy essentials and whatever is left, they will use it to buy things they don't need. For example, when the 99% turns 18 (or 21), the first thing they buy is a brand new car! They don't realize that this shiny brand new car will lose 30% of its value the very moment they turn on the key for the first time. After this smart move, they will buy a house, paying the mortgage with their salary. They will be locked to work for their bank for the next 30 years, making themselves de facto slaves. The 99% always buys liabilities and never buys assets which can generate passive income. Remember when we talked about Tom and Jessica. Luckily, even this trend is slowly changing, though not fast enough. We hear more and more about passive income, but still few people know what passive income really is and fewer put the concept at work. Passive income is any stream of money getting into your account without your active, constant work, in a passive way, indeed. Usually you work on it only once and after the initial investment or setup, money will flow to you automatically. A classic example is real estate. You invest to buy a property, you work your ass off once to fill in all the papers, but once you rent that property, you just need to wait for your tenant to pay you each month. You don't have to go yourself to the property every day to carry a task, in order to generate income. Another classic way is to buy stocks in a company, which entitle you to receive a slice of the company's revenue (dividend). You will get your dividend even if you don't work in that company. The Internet, like usual, brought a massive amount of new ways of generating a passive income, but also a massive amount of scams and disinformation. For sure you've seen a million times ads like "I can't believe when I saw my first check! $3000 in just one night! Let me show you how! Fire your boss today!"

All of these scams leverage your frustration about your job, your life, your income. After all, it is said that 85% of people hate their jobs.[43]

I don't know the secret of that remaining 15%, those who feel engaged at work, but I assume they have some form of passive income, that releases them from the burden of working just for money. At least in my experience, this led me to see something more in what I do daily at the office. Money is just a consequence of the challenging problems I manage to solve. When work is not your only source of income, you begin to be more relaxed, your confidence rises and you start taking braver decisions, which eventually help you get promoted, increasing your responsibility, your salary and hence, your job satisfaction. Maybe you'll never manage to fire your boss with your passive income, but you can still improve your life at your workplace. This book is not about telling you how to create sources of passive income, but the best advice I can give you is to buy the book you should read after this one: *Rich Dad, Poor Dad* by Robert Kiyosaki. This volume is essential to understand how money works and how to make it work for you. Thanks to it, you will start reducing your unnecessary spending, improve your life-style and work to BE rich rather than to LOOK rich. Once you will have read *Rich Dad, Poor Dad*, you will also be able to spot those wankers who try to sell you magical methods "that only the 1% knows" to make money "overnight". For sure they make money overnight, YOUR money...

[43]https://news.gallup.com/opinion/chairman/212045/world-broken-workplace.aspx

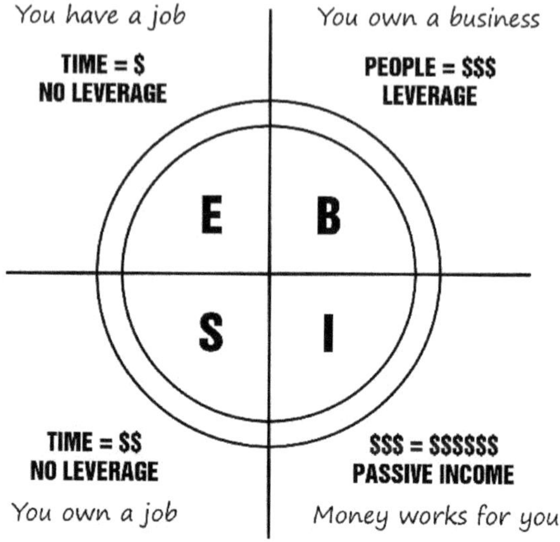

I posted Robert Kiyosaki's quadrant to link financial education to inequality generated by capitalism. In our society there are 4 different main categories of people: employees (E), self-employed (S), business owners (B) and investors (I). Around 99% of the population is either E or S, but holds only 1% of the world's wealth. On the other hand, only 1% of the population is either B or I, but detains 99% of the world's wealth. In Robert Kiyosaki's view, the only effective way to change this unbalanced status quo is to provide serious financial education, teaching investment management since primary school.

With a population financially savvy, probably Jeff Bezos wouldn't own the majority pack of Amazon, and his billions would be spread among more individuals. Companies with multiple mini-stakeholders can exist and thrive too! I used to work for one of them: Flutter Entertainment, which includes brands like PokerStars, FoxBet or PaddyPower, is controlled by several small stakeholders, who elect a CEO and a board of directors. There is no megalomaniac billionaire behind them. To mention something much bigger, you can look at General Electric. GE is estimated to be one of the most colossal companies in the world, with an income reaching $5.23 billion in 2020! The company reached these numbers without any clown performing the stunt of crashing his own truck on stage. Everyone can own a share in GE, even you!

Who knows? Maybe GE will be the first business you've ever owned... You will be a little Elon Musk, but better and more likable! Because our society doesn't need those characters. These billionaires try to sell a distorted view of the American Dream, telling us how they started from nowhere, with nothing, and how they became "self-made" men. There is no one who is self-made. First of all, Musk and Bezos didn't start with nothing, but they had wealthy parents backing their first attempts. Later on, they were backed by VC investors (exactly, those who make money work for them, bravo!). Sure, convincing investors to bet on your project is a quality, but even choosing the right project to invest in is a quality... Indeed, Musk would have become a VC himself, but in his own ego-centric way. You know him mostly as Tesla CEO and co-founder, however... he didn't confound Tesla at all! Tesla Motors was created by the brave minds of Martin Eberhard and Marc Tarpenning, who envisioned a car company which could also be a tech company with the disruptive spirit of a Silicon Valley start-up. Their wonderful dream brought on the market the first real electric car capable of competing with traditional combustion cars. This forced traditional car-makers to finally wake up, after decades spent on delaying the green revolution, even though the technology was already available. Too bad that Musk joined the very first round of funding for the company and, along with the majority of stocks, he even took the merit, building up his mythical image of self-made entrepreneur-scientist. Musk would be nothing without his family's wealth, without the VC investors who believed in his

PayPal project and the brilliant employees he surrounded himself with at Tesla, including the real co-founders.

Here is the third, final take from the billionaires' tales. They convince you that you can become like them, so you keep voting for who supports them and who continually cuts taxes for them and their big corporations. They justify this doctrine with an alleged trickledown effect. *"If you cut taxes for us, the chosen ones, we can create more jobs for you and further boost the economy"*. This can work in principle, but only within certain limits. Do you think it's normal that at the end of the day, in proportion you pay many more taxes than Amazon? Unfortunately for these billionaires and their Demo-Republican associates, a 50 years old study (retrospective of course), run for 18 countries, detected that cutting taxes for the rich has no trickle down effect at all.[44]

This study, titled *The Economic Consequences of Major Tax Cuts for the Rich*, by David Hope and Julian Limberg, concludes that:

"We find that major tax cuts for the rich push up income inequality, as measured by the top 1% share of pre-tax national income. The size of the effect is substantial: on average, each major tax cut results in a rise of 0.8 percentage points in top 1% share of pre-tax national income. The effect holds in both the short and medium term. Turning our attention to economic performance, we find no significant effects of major tax cuts for the rich. More specifically, the trajectories of real GDP per capita and the unemployment rate are unaffected by significant reductions in taxes on the rich in both the short and medium term." [45]

---

[44]https://gulfnews.com/business/banking/fifty-years-of-tax-cuts-for-rich-didnt-trickle-down-study-says-1.1608092865109

[45]http://eprints.lse.ac.uk/107919/1/Hope_economic_consequences_of_major_tax_cuts_published.pdf

This is a shocking breakthrough; bound to shake what we've been believing for so many years. It led to change my mind too. I had always been a supporter of cutting taxes for corporations. The 18 countries observed were:

- United States
- United Kingdom
- Sweden
- New Zealand
- Italy
- Netherlands
- Germany
- Canada
- Australia
- Norway
- Japan
- Ireland
- Finland
- Denmark
- Belgium
- Austria
- Switzerland
- France

As you can see, these are all functional Western democracies, where we always hear the rich demanding more and more discounts on their tax shares, so they can invest more in their communities and employ more people.

Don't get me wrong; I'm not a fan of taxes either and I deem insane the taxation rates of places like Italy, France or Germany. Those are not taxes, but extortion. However, it's also true that too little taxes raise social inequality and make a state ineffective at serving its own citizens. How can you run schools, build roads or assist the handicapped and the old if you don't collect enough taxes? Mostly, how can you keep ICUs up and running...?

We can see where this line leads, if we look at the tax

heavens set in the Caribbean. There, the rich pay like 0 taxes, but the local population starves to death, with no public services available, whatsoever. We can't escape from the law of balance. Balance is always the word number one, in every situation, including taxes. America lost the concept of balance long ago, pushing the nation, decade after decade, to become a tax heaven for the super-rich and a tax hell for the normal people like you. Do you still want the show to go on? I think you can have better... Listen to President Corona. He came to open your eyes and to make you understand that you've been fooled, for too long...

Be Water!

Unlike what conspiracy theorists believe, businesses didn't bend before communism, but found a way to adapt. Business is all about adaptation to new conditions. Only the fittest survives and successful entrepreneurs focus on the solution, not on the problem, to adapt as fast as possible. I know it's a concept hard to grasp for you, because you are used to whining over every issue, focusing on the problem and never on the solution.

Let's take the food business as an example. The stupidest conspiracy theorists even formulated that COVID-19 was a big staging with the main purpose of killing the catering business, so that the state gains control over food supply, paving the way for communism. It didn't take much to debunk this "shrewd" theory, but just in case, we can see how restaurants managed to thrive by changing their approach with services like Bolt Food or Uber Eats.

- Problem: You cannot serve food to people at your venue.
- Solution: Work on delivery.
- New Problem: Your fleet is insufficient to meet the delivery demand.
- New Solution: Outsource delivery to a central platform that will provide you with the delivery guys!

You see? No whining, no waste of time, no waste of energy to overthink about dark master plans, 0% communism, 100% pragmatism and optimism! You don't have the latter? Then you will never succeed in life, regardless of the model adopted by society.

The aforementioned tools came right on time, saving restaurants during the lockdown! Before, many places could rely only on their internal fleets, usually composed of a few scooters, which were incapable of serving more than 10 customers a night. Now, even the smallest kebab kiosk can scale up its business and customers don't have to move their asses to taste frozen Chinese food fried in sunflower oil! Now they can get this shit effortlessly, while waiting for tonight's match comfortably from their couches. From the employment perspective, students found a new hustle to fund their studies with, while migrants found a new entertaining way to be exploited with.

What President Corona Taught Us

Freedom is never granted. Humans fought hard for it and they are aware these efforts are still not enough. We have been enjoying things like equality, civil rights, free speech for so little time. The modern concept of freedom is as young as a couple of centuries and it's only since the WWII aftermath that this concept has been kept alive without interruptions. We must thank President Corona for having challenged us on how to make freedom coexist with security, to understand the border between individual freedom and collective freedom. I'm talking about the right of the individuals to fulfill their own lives versus the right of the entire population to feel safe. This very topic sparked the harshest discussions.

We Must Learn to Listen

The legitimacy of stringent measures to protect public health at the expense of individual rights created the most violent debates. For the defenders of public health, the counterpart was a bunch of selfish, spoiled kids. For the defenders of individual freedom, the counterpart was the front of neo-fascism, eager to strip our freedoms in a sadistic effort.

Both parties have been trying to bend reality to their favor, with distorted numbers, facts and data, just to have the upper hand. We rarely saw constructive debates where the contenders tried to reach common ground. The sad part is that we forgot we all want to live a fulfilling life and we all wish everyone is safe. Have we really tried to reach a common ground? I have met only a couple of examples. One was an old man who supported the first lockdown. He said it was legit, given how little we knew about the virus and how inadequate our hospitals are to face a respiratory pandemic. However, he even conceded that the restrictions should not affect the young too much:

"I'm 78 now. I have lived a wonderful, adventurous life. I've done everything I wanted to. If this virus is bound to catch me, I'll welcome it with open arms. It's not fair to ask the young people to sacrifice their youth so I can live some years more. I can't ask them not to live their lives, when I lived mine to the fullest."

That was simply beautiful, as beautiful, even wise, was what an anti-vax told me:

"I want to specify I'm not against the vaccine itself. I even believe in vaccines. I'm just against imposing it. I also understand we live in a society and that the majority wants the vaccine. Fine, I keep myself from society. I'm not going to the restaurant, I'm not going to the clubs. I don't because it's not fair to impose my battle on others. And in the end, we never know who's right. I believe in my fight, but I cannot be sure 100%."

These are two rare examples of people capable of walking in the other's shoes. These are the people who listen with the intent to understand, not to reply. If we had more of these people, the world would be like heaven!

We Can Learn a New Job!

Our parents used to live in a world where you got a PhD in a given field, started to work in such a field at 24 and finished to work in the same field at 60-65 years of age. All of your life, the same, safe, but boring job. Monday to Friday, 9-5, for the rest of the best days of your life... Holidays only during August, the worst month ever to travel...

Despite the 21st century and the digital revolution brought a much more dynamic world, we still lived in the same mindset our parents passed onto us. We inherited the same fears too. How many friends came to me complaining about their job! This company sucks, this job is boring, I'm not growing and blabla... But when I asked them if they were looking for something else, fear kicked in: "Are you crazy??? I can't leave them! - It's a good environment after all. - When you leave, you don't know what you may find! - Another company might be even worse! - In the end, here I can still do whatever I want..."

All excuses! They hate their jobs, but they never had the balls to leave. They leave only when it's too late. When they are laid off! When they are forced to! Those who never face the doomsday, are bound to live unhappily, licking the boots of their bosses and pretending that everything is ok, at the end of the day. This attitude is sad. *The Matrix* is just a metaphor to describe this! Indeed, Neo is a white collar unsatisfied about his working life, confined in an anonymous cubicle, abused by his boss. It's not about machines which created a simulated reality where to sedate the humans they use as batteries! That's just the entertaining layer. *The Matrix* is about releasing yourself from the reality you have created about your own life, thinking that everything in your life-path is already decided from the start, that you can't escape your destiny, that dreaming about something else is stupid, useless or even wrong! Luckily for most of you, in March 2020, President Corona knocked at your door, but instead of sending you into the ICU, he offered you the red pill!

President Corona's red pill shows you how the importance you gave to your job was an illusion. For one job you cover, there are hundreds of better roles waiting for you. All it takes is courage. You're not skilled for them? You may want to rephrase this statement to: "I'm not skilled for this job YET...".

I'd like to highlight a laudable initiative from Microsoft. The company started a project to help 25 million job seekers to find a job in a Covid-19 world. They created the portal called https://opportunity.linkedin.com/en-us in collaboration with LinkedIn. Here you can find the path you wish to take, in order to change your career and live a new life as a worker qualified and skilled for this new world. This is just one of the several tools you have nowadays to reinvent yourself. There is no reason to sit down and cry. There are all the reasons to stand up, study, practice and then smile when you sign your new job contract!

Take note of what Microsoft's CEO Satya Ndella says:

"We believe that we need a culture founded in a growth mindset. It starts with a belief that everyone can grow and develop; that potential is nurtured, not predetermined; and that anyone can change their mindset. It's not by claiming a growth mindset but by knowing that we are imperfect but can learn and get better." [46]

Similar initiatives followed soon after. Bumble, the popular dating app, launched a program which offers women free courses to become developers, with the promise of being hired by the company once they graduate! The program is called Bumble Tech Academy, developed with tech school CodeOp. The campus is to be held in Barcelona. For now, the campus is open to women and non-binary with the right to work in the EU, but I'm sure the company plans to spread this initiative to other countries too. The courses are not only free, but the lucky students shortlisted for this training will receive

---

[46] https://www.linkedin.com/pulse/our-opportunity-define-world-we-want-live-satya-nadella/

€25,000 a year during their training period! Not bad!

https://codeop.tech/bumble-tech-academy/

This is a great trend that is set to grow. I see more and more companies training their future employees. It is a great idea which benefits both parties. A student can land a qualified job without having to invest upfront for the required education, without the worry of having to look for an employer at the end of the academic period. In return, the company is going to hire someone who is already known and trusted, skilled in what exactly the business needs!

If you can't make the final cut for projects like this, fear not my friend! There are plenty of platforms which allow you to acquire the hottest skills required by the job market. Most of them come very cheap and some come even for free!

The pandemic could be a great occasion to enhance your skills. If you want to acquire a high demanded specific skill within a short time, in my opinion the best tools are:

- LinkedIn Learning (first month is FREE)
- Udemy
- CodeAcademy
- Master Class (if you want to have the fancy experience to learn from the stars of your industry)

If you aim higher, for example to get a master degree, comfortably at home, you may want to check Coursera. They offer courses from the most prominent universities in the world, including lectures. You can even follow most of the courses for free and pay only if you want the official certification with a related degree. Thank goodness, we have infinite possibilities today. It may sound ironic, but there has never been a better time to be unemployed! Should the pandemic have happened, for example in the 1980s, probably I wouldn't have been so optimistic. Think how lucky you are to live in the 21st century...

If you think I'm talking out of arrogance, well I must tell you that I actually left my job too, while I was writing this book.

One sunny day in July 2021, I called my manager Kate and told her I was going to resign. At first, Kate believed I was joking, then while explaining my reasons, we both started to cry. Kate was so empathetic that she made sure I was sane while taking this step: "Wait a moment. Do you have enough money to leave your job without anything else? Please, don't rush!"

Well, I had money. Not only my savings, but also my crypto, especially the ones I had the courage to sell during the bull run. Do you know what's the best thing I've ever done with crypto? I spent them!

Kate invited me to cool down for a few days, to think carefully about my reckless intention: "Let's have a meeting again on Monday. If you still have the same idea, then I will accept your resignation."

After the call with Kate, I was devastated. Resigning from a company you love feels as painful as a break-up, I'm not exaggerating! When I hung up, I was unable to work again. I spent the rest of the day lying on my bed, without a single bit of energy. I even experienced physical pain, as if all of my bones were breaking at the same time! Just as it happens when you break up with someone you're still in love with!

The next Monday, my idea wasn't any different, but I was calmer. Deep inside me, I felt that was the right call. All I could do was to send an email to the HR, my team leader Dmitri, my manager Kate and all my fantastic work buddies.

****

Hello Team,

I hope you are having a great time and that you are safe!

It's with great sadness that I have to communicate my need to resign from my position.

When you work for PokerStars, you must give your best every day, in every situation, to every customer and every colleague.

I can guarantee this only until July 18th 2021.

Taking this decision wasn't easy at all. You can't leave 4.5 years behind like nothing. I confess that I shed more than a tear during these days. PokerStars is by far the best company I've ever worked for, the place I've always dreamt to be since when I started playing poker.

I plan to spend the upcoming months travelling, learning to enjoy life more, rebooting my brain, requalifying, volunteering and taking care of my mental health. Who knows, maybe my new future skills will allow me to rejoin the family soon, under a new form...

In any case, I will keep my stake in Flutter, so technically, I will still be your boss!

I'm sure that you will make me proud and that this ship will keep sailing steady through the Seven Seas! The wheel is in good hands. I have total confidence in Dmitri's skills. You will be a great Team Leader!

A big hug,

****

When you choose a job, make sure to work with great human beings. I will never be grateful enough toward the peers I had the priviilidge to work with at PokerStars. They simply rock!

After handing down my work duties, I spent my crypto traveling around Europe to experience life again, after 1.5 years stuck in Malta, while also searching for a new home. It was the bravest, yet the wisest decision of my life.

It can be hard, but you can do the same! I can already imagine you in your new job. I imagine you'll be scared like a kid during the first day of school, but at the end of that very first day, you will feel confident and happy!

We Can Change Partners!

Yeah, your ex... How many people hate their romantic partners, but nevertheless, they remain bound to them for the rest of their lives? How many times we heard: "This time is over! I'm bumping you!", but the next day everything is still the same? We are so afraid of change that we prefer to keep on suffering rather than finding an escape. Most people need a push to quit their failing relationships. Most of the time, that push never comes and only death manages to separate the lovers (or haters?). President Corona offered the chance to many couples to get the push they would have never got in their lives. He broke up many toxic relationships, for good! When you are trapped in the same house 24/7 with a person you don't like, you will run out of excuses very fast.

British law firm Stewarts registered a staggering 122% increase in divorce enquiries between July and October 2020, compared to the same period of 2019. This data matches with the ease of lockdown in the UK. As soon as people were free to go out again, separations went to the moon. The lockdown has been a powerful generator for divorces. It's just in the moments when families are together for long periods, like during holidays, that they burst out. It's no surprise that most divorces are filed right after Christmas... The lockdowns offered a longer period also to rethink about our existence. We understood that life is too short to spend it with a person we don't like. President Corona's message is that we should not rush into a marriage or any relationship if we don't really want to.

It's time to drop the idea that if you're not married you're a loser, as if marriage was the the only way to be truly happy. This is just a horrible misunderstanding, an idea carved in our minds by Disney's movies. If your happiness still rely on the person you happen to be by your side, I suggest you to read Eckart Tolle's book *The Power of Now*. Tolle reveals that true happiness can come only from within you. Until you won't be happy on your own, you will never be truly happy with anyone else. You may feel happy for a while, especially during the first year of a romantic relationship, when chemical love acts on your brain like a drug. Once chemical love passed, what's left are two strangers forced to live together and "love" each other to make sense of their commitment, to have a wing in life and to satisfy their basic needs. There will be nice moments, alternated by glimpses of hate, which with time may lead to stress, emotional pain, even violence and finally, to a dramatic break-up...

Here, the President's merit is that he just saved a lot of time and pain for many couples around the world, by speeding up the end of their bond. I know President Corona's methods can be quite brutal. His philosophy is: *better a great pain today, than a small pain for a lifetime*.

However, we should not forget the huge collateral effect of this policy. The lockdown inflicted a really big pain to those victims of domestic violence. When your partner is abusive and you are forced to live with him (sometimes even her), life can be hellish. The impact of the lockdown measures was devastating for the victims of domestic violence. Research suggests that domestic violence cases rose steadily during the lockdown, but reports dropped. This is expected, unfortunately. When you are confined with your abuser, you are terrified that he might find out you have reported him. You are afraid for you and your children, another category dangerously exposed to domestic violence and abuse during the lockdown.[47] [48]

A smart strategy developed during the COVID-19 pandemic is the so-called #SignalForHelp. It is a simple hand gesture. You raise your palm in front of you, tucking in your thumb, then lowering your fingers to cover it. This was started by the Women's Funding Network, as a coded, discreet, safe gesture that victims can use during video calls, for example. During

---

[47]https://pubmed.ncbi.nlm.nih.gov/34402325/
[48]https://pubmed.ncbi.nlm.nih.gov/33006492/

the lockdown, video calls have been the only form of contact with the external world and hence, the only way someone could ask for help.

The best way to use this signal is to follow the Reach Out, Listen, Respond pattern[49]. As stated on WFN website:[50]

- REACH OUT: Use another form of communication (SMS, WhatsApp, social media, email)
- LISTEN: Ask YES or NO questions.
- RESPOND: Only call 911 if the survivor asks. Let the survivor tell you what they need, how you can help.

What might be the most important innovation during the COVID-19 era, is not a revolutionizing app, nor a new cryptocurrency, nor a new military drone, but a simple gesture conceived by empathetic people who were questioning how to save lives from domestic abuse. All the technology we can develop will never save us as much as empathy can.

We Can Live Without Watching Sport

When the Premier League 2019/2020 was halted due to the pandemic, Liverpool FC fans almost had a heart attack. They had been waiting for 30 years to celebrate the national title. They were leading the table with an impressive pace, which left every other opponent light years behind. The Reds were projected to end the season with 100 points, but on 14th March 2020, the Football Association deemed opportune to halt the season. Too many players and staff got the virus. It was the second time in history that English football halted a season. The first time was during the 1939/40 season, halted due to WWII. Liverpool just needed 3 points more to reach the title. It felt like COVID-19 teased the Reds. They got so close to touch what they had been longing for so long. It felt like a curse.

---

[49]https://www.domesticshelters.org/articles/escaping-violence/how-to-spot-a-signal-for-help-and-how-to-respond

[50]https://www.womensfundingnetwork.org/signalforhelp/

Maybe COVID-19 happened just to keep Liverpool away from the national title!

The sadistic British tabloids began to torment Liverpool fans with speculations about a possible outright cancellation of the 2019/20 season, which would have nullified the groundbreaking records set by Jurgen Klopp's team over the season. The Reds could even make 200 points. They will never win a Premier League! These speculations would have haunted Liverpool fans from March 2020 until June 2020, when the matches were resumed, though behind closed doors. That was enough to allow Liverpool to secure the title of English champions, 30 years after the last honor won in 1990. The joy of the Red fans was so great that their celebrations caused a massive spike of Covid cases in Liverpool, forcing the city to undergo a new lockdown in summer 2020. The hottest lockdown ever!

Meanwhile, people all over Europe realized life without football wasn't that bad. The rest of the world followed a similar pattern. What's life without baseball for an American? What's life without cricket for an Indian? What's life without rugby for an Australian? Maybe not so bad. People were finally free from the ritual of watching their team playing live, no matter what. It's scary to think how watching sport can enslave you. If movies and TV shows give you the flexibility of watching them whenever you want, sport is a tyrant. There is no point in watching a game when you already know the result. Sport is served religiously live. This can condition your life heavily. When I used to live in Italy, I gave up on many social interactions due to sports commitments. Formula 1 was my Sunday Mass. I even used to wake up at 5.00 am to watch a GP taking place in another time zone and I always rebuffed my friends' offers to spend an outdoor Sunday if this clashed with a F1 race. The most important football games were a must too. If I could give up on the minor games, I had to follow the great classic games like Barcelona – Real Madrid, Bayern Munich – Borussia Dortmund, Liverpool – Manchester United, Chelsea –

Manchester City or AC Milan - Inter (it's just Inter, not Inter Milan!). I'm mentioning a few, but they were many more and the Champions League was another unmissable appointment. When I moved to Malta though, something changed. I discovered life offers better activities than watching 22 millionaires chasing a ball. F1 became a distant memory too. I can't even remember the last race I watched. I still watch some football games from time to time, but still in a social environment and without commitment.

Malta was for me, what Covid-19 was for the other sport aficionados around the world. They could finally wake up from the sport matrix and see all the plethora of alternative activities they could do besides watching sport. For sure, all those wives not into sport, must have thanked President Corona for creating extra time to spend with their husbands. For those wives who are into sport as well, they might have discovered there are many other great activities you can do with your partner when sport is not on. Some finally started to practice sports. Physical activity is one of the few things local governments allowed to do outdoors during the lockdown. This led many to realize that sport is better when done. Sport is better when it's about your own wellbeing rather than about sick competition.

What I'm glad about the pandemic is how hard it hit the sick European football world. I began to feel pure nausea when I read how players are transacted for insane millionaire fees and how their salaries were going beyond any reality. I understand a footballer must earn a lot. To reach the top you must undergo incredible sacrifices. Talent alone is not enough. Forget to live your teenage years like the other kids. You must be a soldier. Beside sacrifices, footballers are the ones bringing millions into football, so they deserve a big chunk of such money.

However, footballers should not forget that such money comes from the very fans. Many fans don't even see €2,000 in a month. Players should be more considerate when negotiating a new contract and have more touch before skipping to a rival team in the face of their fans who supported them.

On the other hand, fans are at fault too. They complain about the insatiable demands the players and their managers have, but at the same time, fans are likewise demanding with their clubs when it comes to setting up a competitive team capable of winning trophies. Fans demand higher and higher expenditures to secure the best players. That's how players became objects of real auctions. One day we will see Kylian Mbappe for sale on eBay... It's a vicious cycle built from people who think the toy will never break. Then one day a tiny virus, an invisible leader brought them to reality. That was the first time I saw him. I remember a day of 2020 when football clubs were already dealing transfers for the next season, amidst the pandemic. Suddenly, the figures were so lower than the previous summer. Clubs were no longer the same braggarts capable of shooting €100 million for the newest marvelous talent. One of the most lavish clubs, Paris Saint Germain, accepted to buy Mauro Icardi from Inter. My dad is a huge Inter fan and was still very attached to Icardi, despite his bad behavior during his last months in Milan. My dad complained the transfer fee was too low, "just" €70 million. I texted him back with a revelation:

"The fee is fair. The era of €100 million players is over. Another great decree of President Corona!"

That was the day I realized this wasn't a normal virus. This wasn't just a pandemic. This was a revolution. The wind of change was blowing. The world was going to be led by the first president of the world. That was the day I saw the disruptive, but revolutionary agenda of President Corona!

Nations Are Imaginary Borders

"This is a Chinese virus!"

– President Donald Trump

No Donald, a virus doesn't have nationality and doesn't know any borders. The only border a virus knows are species. Before he used to live in bats, then he jumped on humans. Because the border between species is tangible. A monkey and a turtle are really different. Their biological processes change dramatically, hence their needs: what they eat, where they can live, how they reproduce. If you take an American and a Chinese, their biological processes are the same, hence their needs. They are both humans (even if some scumbags might not agree). They need the same kind of food, they need the same temperature range to live, they need the same demanding comforts humans do and yes, they reproduce in the same way. The only difference between them is given by an imaginary border, drawn in the collective mind of humanity, which says one is American and the other one is Chinese. The concept of nation itself is an illusion, so ethereal that indeed nations change their shape over the centuries. Today we like to think that modern nations inherit the culture, the history or even the genetics of the people who lived on their same lands centuries or even millennia before. Nothing more distant from the truth. Let's take Italy for example, which Mussolini considered the legacy of the great Roman Empire. Roman culture influenced Italian culture heavily, but not only the Italian one. It influenced the whole European civilization and beyond. Italy's history begins longer after the fall of the Roman Empire. For centuries, Italy was an agglomerate of different mini-states, most of the time at war. Wars were even fought between cities.

The idea of a united Italy came up only in the 19th century, mostly brought by the changing wind of a "foreign" invader. Napoleon created a republic in Northern Italy. The project was short-lived and was wiped out by the restoration, but locals began to believe they could be one nation. This idea was translated into reality 40 years later, by people who were completely different from the Romans. They dressed differently, they spoke another language, their thoughts, hopes and dreams were more complex, projecting well beyond simple survival. Even their genetics, which some still deem important to define a nation, were different from the Romans. Over the centuries, their blood was mixed with Arabs, Greeks, Turks, Germans, French, Swedish and more. This was possible, because every human is genetically compatible with any other one. Because the human species has only one race, one breed. There is no human counterpart for a German Shepherd, a Husky or a Beagle. You can talk about ethnicities, to describe certain physical traits, given by specific genetic combinations, which are called phenotypes. These phenotypes might rise in a certain geographical region, giving peculiar physical characteristics to a determined population, but they are meant to spread and mix with the others, not to remain in a virtual cage to preserve a certain "race" or to avoid "contamination" by another one.[51]

Not even evident ethnic traits are sufficiently tangible to separate humans into distinct races. The same can be said about nations. If you could go into space, you would see how

---

[51] https://www.psychologytoday.com/intl/blog/looking-in-the-cultural-mirror/201109/the-main-reason-races-don-t-exist

there are no real borders between nations. Only checkpoints invented by humans, but Earth is not born with them. Its biosphere is not aware of these lines. An animal of the Amazon will never feel Brazilian or Bolivian. To assess this, you don't even have to go to space. You can go on earth.google.com. From here, go to the left bar, click on the map style icon and choose "Clean" (no borders, labels, places or roads). With this option, you can admire Earth as if you were in space; clean indeed. Clean from our imaginations for which we keep on fighting stupid wars and putting non-sense restrictions.

Let's think about, for example, the United States of America. They are said to exist in the North American continent and that they are very powerful, advanced and wealthy. Who says that? All the humans on Earth. Why? Because they agreed that in a region which stretches from the Atlantic Ocean to the Pacific Ocean, extends till the Ontario lake to the North and the Sonoran desert to the South, there is a nation which is called United States of America, where people communicate through a system of vocal sounds called English and abide to a complex series of rules called American laws, which in turn, stem from a code of supreme rules called Constitution of the United States of America. These things might be represented into physical supports, but they are ultimately born from our minds. They are a powerful collective illusion, as everything embedded in our society is. If tomorrow, every human stops thinking that the USA exists, this nation will stop to exist. They may decide to call it something else, like Dreamland, change its size, its currency or even its language. If enough people would believe that, the change would happen overnight.

Now, the fact that the whole society we built is fiction, doesn't mean it's useless. Even if society is an illusion, its effects are very real. For example, because of American law, people are expected to conduct business according to strict principles of honesty and integrity. Who commits fraud is punished with fines or even with deprivation of freedom. They are expected not to harm anyone physically. Again, who harms another human, is normally deprived of freedom. Nations likewise, have been a powerful tool which helped humans have a view broader than the small tribe, the setup our brain evolved in. Nations motivated humans to pile up together resources exponentially bigger than what was possible in a small village and this prompted the construction of incredible structures no other animal is capable of. But, as we say, they are still tools. As such, we should use nations and not let nations use us. Why should we swear undying loyalty to a nation even when it wants to lead us into a bloody war? Why should we give up our freedom of movement just because in some part of OUR planet, our national passport is not accepted? Why should we waste our limited, thus precious, time to argue about which nation is the best? Mostly, why should we think that our nationality defines our personal qualities and flaws? You're not good or bad because of what your country mates did, but because of what YOU did, with your own life! You are not your nation, remember that! Celebrating a victory or crying for a defeat of your national team is pointless!

President Corona made this concept so clear. While governments were desperate to protect their borders, the President was crossing them with disarming ease. Because he knows national borders don't exist. He lives in the real world and the real world doesn't have borders. Nevertheless, governments kept on failing to understand this. From the beginning, governments tried to outbid other countries for medical equipment such as masks and respirators, thinking that, if they could get them before anyone else, their people would be safe. The same script was rehearsed during the vaccine roll-out. Again, the wealthiest countries outbid the poorest in the race to stockpile doses. They didn't understand that nobody will be safe from the virus, until every human in the world is safe. If only a portion of the world is vaccinated, the virus will keep on evolving among the unvaccinated half of the planet. Once ready to escape the vaccine, the virus will return to the vaccinated world and governments will be brought back to square one... Do you understand that in a pandemic, nationalism is useless? Put down your stupid flag and do something to help the world and your race, the human race!

Our Enemies Are Not Other Humans

Among the ranting considerations made on this crisis, one gave me inspiration for this chapter. I was talking about Boris Johnson with my friend Simone and he said it was a shame how much the UK spent for the military compared to the NHS. It sounded like the Brits spent much more for weapons than on hospitals. The answer I found for 2019 was that the UK spent £42.2 billion on defense. To keep the British people healthy, Downing Street put £129 billion. Luckily, the good old Boris spent more for hospitals than for tanks, but the money spent on weapons is still insanely high, too high at a first glance.[52] [53]

How are the other countries doing? Are they also spending too much for their armies and not enough for their NHS? Let's take a look at this chart prepared by Statista which shows the military expenditure of each country in % of its own GDP. This is a measure that tells us how "aggressive" and militarized that country is, better than absolute figures.[54]

---

[52]https://www.statista.com/statistics/298490/defense-spending-united-kingdom-uk/
[53]https://fullfact.org/health/spending-english-nhs/
[54]https://www.statista.com/statistics/266892/military-expenditure-as-percentage-of-gdp-in-highest-spending-countries/

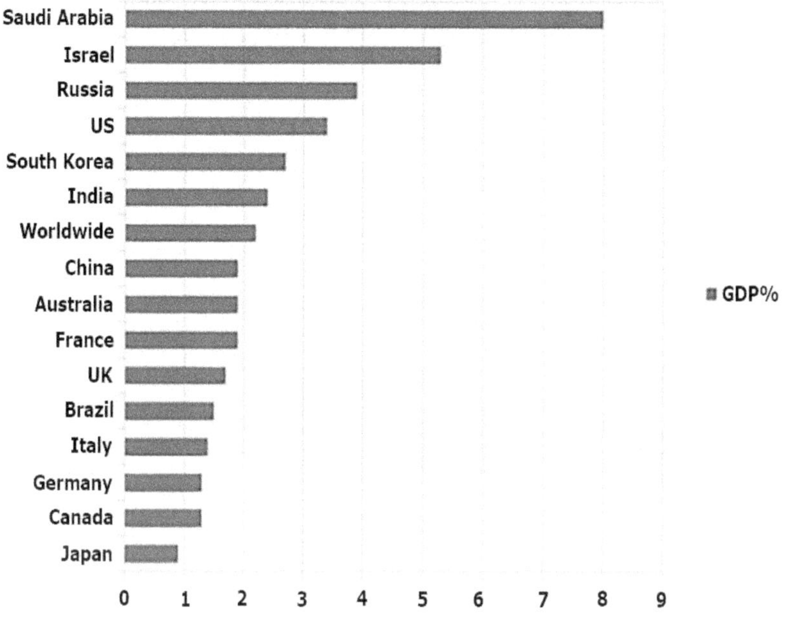

As you can see, it's insane how the Saudi care about their arsenal. They pledge 8% of their national GDP. As predictable, Israel also goes beyond 5% of its GDP to secure its borders.

These numbers are staggering and when you visualize them in absolute terms, more questions arise. The biggest spender in absolute terms, as you can imagine, is the US with a massive $732 billion spent in 2019.[55] That's like 3 times more than the second biggest spender, i.e. "evil" China with $261 billion put into its armed forces during 2019. Even the most classic enemy of America, Russia spent "only" $65.1 billion for the same year. The rest of the biggest spenders are neutral (India) or allied to the US like France, Germany, UK, Italy or even Saudi Arabia itself. President Corona would like to know

[55]https://www.statista.com/statistics/262742/countries-with-the-highest-military-spending/

if it really makes sense for the US to spend all of this massive amount of money on their behemoth military apparatus when no other power can realistically threaten America's borders and sovereignty. Yet the average American will reply: "Without these expenses we would have no free country!". When you have a propaganda machine that works like a clock!

How much do the US and other developed countries spend on healthcare? At a first look the situation doesn't look so desperate.

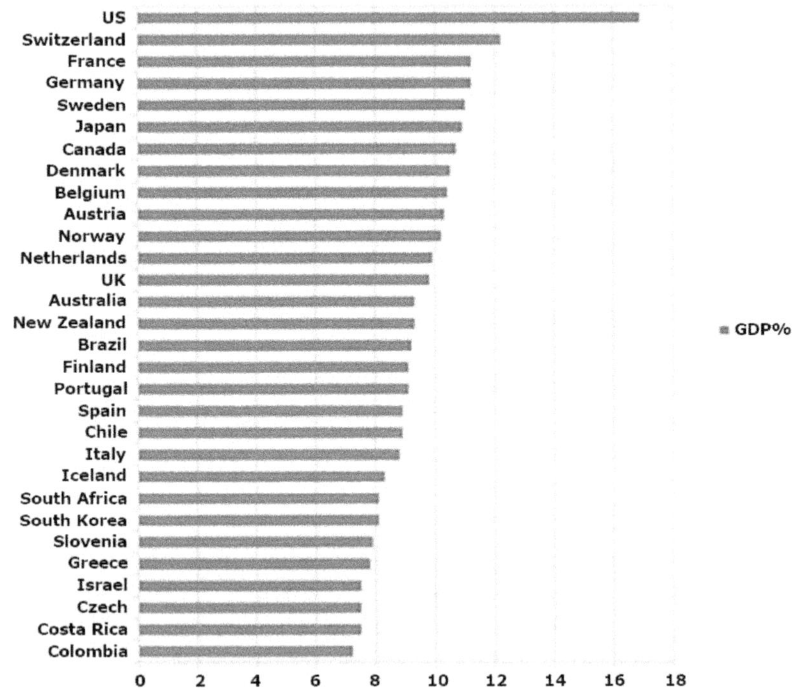

Health expenditure as a percentage of gross domestic product (GDP) in selected countries in 2022[56]

The US spend 17.2% of their GDP on healthcare! They spend

[56]https://www.statista.com/statistics/268826/health-expenditure-as-gdp-percentage-in-oecd-countries/

more than the European states, universally deemed to have the best public healthcare in the world. However, before jumping to easy bright conclusions, we have to bear in mind these statistics include even the private healthcare spending as well as the purchase of prescription drugs. As you can already imagine, these two amounts are by far higher in the US than in any other country.[57]

Finding the absolute figure of public health spending is not easy, but according to taxpolicycenter.org, the American government actually spent $1.2 trillion in the fiscal year 2019 to protect the health of its citizens. This amount equals to circa 5.6% of the American GDP, about $500 billion more than what spent to protect the country from other humans (military spending). While indeed healthcare spending can be either private or public, military spending is only public and here is the fun part! Any time you talk about public spending in the US you are immediately labeled as a "communist", but nobody calls you like that if you propose to increase the budget for Uncle Sam's army. This is the demonstration that American nationalism is like a blind fold to hide from its citizens the biggest paradox of the American nation: the American armed forces are the biggest communist organization in the world, a true Red Army which extorts $732 billion a year from ALL the hard working American taxpayers!

The President made us more citizens of the world. He demonstrated how our national borders are just imaginary. They are not real. A nurse or a doctor fighting the pandemic in the front line in a French hospital is more French than who's simply born in France. What makes you more important for the country where you were born than someone who really helps that community? Nothing. Your nationality is not important because it's just about luck. It's just a fucking casino game, mate! Passports should be given more often based on

[57]https://www.forbes.com/sites/niallmccarthy/2019/08/08/how-us-healthcare-spending-per-capita-compares-with-other-countries-infographic/?sh=45d2ceb575dc

meritocracy. Thanks to President Corona, this happened in France, where non-EU nurses and doctors engaged in the front line received the French passport as a sign of gratitude for defending their adoptive country![58] An example that other nations should follow suit. Everybody knows how difficult it is to get a French passport, even when you truly deserve it. President Corona sped up the process like no other French president could ever do! Thank You President Corona!

He Prepared Us to Face Future Pandemics!

You must know this. President Corona wasn't the real enemy. He was here to prepare us to face the Emperor. This new pathogen will cause real disruption, will take the life of your loved ones, your friends, most probably even yourself. That will happen for sure, if you will act and think like you did during this pandemic. It's not a matter of IF, it's a matter of WHEN. I'm sorry to be a prophet of doom, but that's what it is. Institutions are already working on it, but the budget is limited and the leaders willing to take heed to the problem are still very few. The scariest part is that the Emperor already peered out in the past.

In May 1997, then still British Hong Kong woke up at the Emperor's call. He made humanity aware of his presence, by infecting 18 humans and killing 6. The Emperor loomed over humanity again in 2003, and since then, he managed to infect 861 people, killing 455... Of all the humans he has infected so far, more than 52% died! You know that SARS-CoV-2 kills just 2% of the infected, right? Work it out yourself...[59]

The Emperor is no other than H5N1, a lethal flu strain which causes the so-called highly pathogenic avian influenza (HPAI A(H5N1)). This strain showed on several occasions to have the ability of jumping on humans. How come that we are not dead

---

58 https://www.bbc.com/news/world-europe-55423257

59 https://www.who.int/influenza/human_animal_interface/2020_MAY_tableH5N1.pdf

yet? Because H5N1 cannot spread from human to human, not yet at least... His natural reserves are birds. He dwells mostly in wild birds, but also chickens and ducks serve as perfect hosts for this virus. This avian influenza kills millions of captive birds each year. During 2014-2015 alone, HAPI H5N1 caused $3 billion in losses to the American poultry industry alone.[60] [61]

Of course, chickens and ducks have more chances to transmit the virus to humans. The problem doesn't lie in the fact that humans eat these birds. Consumption itself is not risky, as cooking can kill any virus. The problem is that the meat industry forces humans to handle and get in close contact with these animals.

Farming is not the only problem, though. H5N1 derives from wild birds and wild birds migrate annually. Their range spans throughout the whole globe. Hence why H5N1 is already a global pandemic among wild birds. Thus, it is important that poultry farms prevent any possible direct contact between wild birds and their poultry. Preventing indirect contact is essential too. Indirect contact happens mostly by bringing into the farm contaminated material, such as boots and clothing dirty with wild birds secretions or feathers. So, if you're a hunter, better not to step into a poultry farm.

H5N1 cases have been detected in poultry spreading the whole Eurasian continent: from China to Denmark, from India to Russia. Since 2008, the Food and Agriculture Organization (FAO) reported that the newest strains of H5N1 are getting more pathogenic, meaning more aggressive. H5N1 mutates very quickly, much quicker than SARS-CoV-2. As of now, the virus' spread in poultry is being contained through vaccination. I know you don't like this idea, but vaccines are working wonders against this threat. It's not the vaccine's fault if you're

---

[60] https://www.usgs.gov/centers/nwhc/science/avian-influenza?qt-science_center_objects=0#qt-science_center_objects

[61] https://www.cdc.gov/flu/pdf/avianflu/avian-flu-transmission.pdf

a superstitious troglodyte. Maybe, knowing that most chicken produced in the world is vaccinated against H5N1 may lead you to drop its consumption. You will make vegans happy!

The most common H5N1 vaccine is produced by Harrisvaccines and it targets chickens. Every adult chicken is injected with one 0.5 ml dose. It has to be stored at -80°C if not used within 100 days. If used within this range, it can be stored at 4°C.[62]

And what about humans? Is there any proto-vaccine ready for a possible spill-over to our species? You might be surprised, but... there is! Sanofi lost the race for a COVID-19 vaccine, but it is far ahead of the competition when it comes to Avian Influenza. As reported by the FDA:

"On April 17, 2007, FDA licensed the first vaccine in the United States for the prevention of H5N1 influenza, commonly referred to as avian influenza or "bird flu". This inactivated influenza virus vaccine is for use in people 18 through 64 years of age who are at increased risk of exposure to the H5N1 influenza virus subtype contained in the vaccine. This vaccine is derived from the A/Vietnam/1203/2004 influenza virus. The vaccine is manufactured by Sanofi Pasteur Inc. of Swiftwater, PA and has been purchased by the federal government for inclusion within the Nation's National Stockpile."[63]

A human vaccine against H5N1 is available in the US National Stockpile, should the apocalypse come. The UK, France, Canada and Australia have similar reserves too. This is great news, but it is not enough to sleep between two pillows. First of all, we cannot take for granted that the first H5N1 strain capable of transmitting from human to human will be compatible with the Sanofi vaccine we have available now.

[62]https://www.drugs.com/vet/avian-influenza-vaccine-rna.html
[63]https://www.fda.gov/vaccines-blood-biologics/vaccines/h5n1-influenza-virus-vaccine-manufactured-sanofi-pasteur-inc-questions-and-answers

Maybe yes, maybe not. It depends mostly on the proteins it needs to spread among humans.

Secondly, there are no similar stockpiles in the regions where human cases were recorded, above all Asian countries. A proto-vaccine would be effective only if administered to the very first people coming into contact with the virus. It would be the only way to avoid the virus spreading to the whole world. It's something we couldn't afford. Remember, H5N1 mutates much faster than SARS-CoV-2, so once the virus leaves the ground zero area, he will take a few weeks to escape from the vaccine. However, President Corona helped us accomplish the greatest technological advancement in preventive medicine ever: RNA vaccines. The coolest thing about RNA vaccines is that they can be edited quickly. Every time a virus mutates, you can have a vaccine ready a few days later. Sure, even this feature has a limit. If a virus escapes a vaccine every month, it's not realistic to come with a new vaccine every month too. No problem from a technological side, but logistically, it would become impossible to upgrade quickly enough the massive production chains a pandemic vaccine requires. We could see already with COVID-19 what a logistical nightmare is to deliver a vaccine to the four corners of the world.

The half good news is that H5N1 responds to some already available drugs. Roche's Tamiflu, used also for normal influenza, showed to be effective in stopping the spread of the virus inside the host's body. For this reason, in 2009, Roche decided to donate 5.65 million Tamiflu doses to the WHO. The donation is not the first of its kind. Roche had already donated Tamiflu doses to the WHO, but these went into the hasty and exaggerated response to the H1N1 swine flu pandemic.

"The recent outbreak of influenza A (H1N1) shows that such a virus can be totally unexpected and spread rapidly around the globe."

– William M. Burns, Head of Roche's Pharmaceuticals

5 million doses come in the normal packet and should be enough for a whole treatment course. The 650,000 are for use on children.[64]

Not that this donation necessarily comes out of genuine generosity. Don't forget that it was a Roche representative who pushed the WHO to declare the swine flu "a deadly pandemic"... Anyway, the latest studies found strains of H5N1 which developed resistance to Tamiflu. This bastard is highly adaptable.

A dire future waiting for us

Who else can H5N1 infect? H5N1 lives in wild birds, with a keen interest on waterfowls, wild ducks, geese, swans and pigeons. Scientists proved in a controversial lab experiment that H5N1 can spread between mammals too, in a few mutations. A team led by Ron Fouchier, conducted the experiment at the Erasmus Medical Center in Rotterdam, the Netherlands. The abstract warns us:

"Highly pathogenic avian influenza A/H5N1 virus can cause morbidity and mortality in humans but thus far has not acquired the ability to be transmitted by aerosol or respiratory droplet ("airborne transmission") between humans. To address the concern that the virus could acquire this ability under natural conditions, we genetically modified A/H5N1 virus by site-directed mutagenesis and subsequent serial passage in ferrets. The genetically modified A/H5N1 virus acquired mutations during passage in ferrets, ultimately becoming airborne transmissible in ferrets. None of the recipient ferrets died after airborne infection with the mutant A/H5N1 viruses. Four amino acid substitutions in the host receptor-binding protein hemagglutinin, and one in the polymerase complex protein basic polymerase 2, were consistently present in airborne-transmitted viruses. The transmissible viruses were

[64] https://www.ctvnews.ca/roche-donates-5-65m-anti-viral-med-packs-to-who-1.397925

sensitive to the antiviral drug oseltamivir and reacted well with antisera raised against H5 influenza vaccine strains. Thus, avian A/H5N1 influenza viruses can acquire the capacity for airborne transmission between mammals without recombination in an intermediate host and therefore constitute a risk for human pandemic influenza."[65]

These shocking conclusions were revealed in 2011 during a conference in my lovin' Malta and could be published only in 2012, after months of heated debate, involving even state authorities. The experiment was one of the most controversial ones of the 21st century. It brought to the table once again the never-ending debate about gain of function and dual use research. Is it worth developing a highly dangerous pathogen to learn from it and tackle the threat in advance or should we bury any knowledge that would make bioterrorism easy? The question rose the upper floors in the Netherlands, but also in America. The US, through their agency National Science Advisory Board for Biosecurity (NSABB) advised against the publication, in fear that the paper could become a cookbook for terrorists eager to create biological weapons of mass destruction. However, after consulting with the WHO, the NSABB changed their position and just asked to remove the very technical steps used to create this strain. If the US was OK with it, Fouchier's homeland couldn't sleep over it. The Dutch government stated that publishing such a paper was equal to exporting weapons abroad. It required a special permit as it is outlined by the EU directive 428/2009 on dual use material. The Dutch government position might seem over the top, but if we look at this strain of H5N1 as a potential weapon, then it makes sense. By publishing the "recipe", malevolent scientists could repeat the experiment on humans, until they obtain a strain transmissible within the human species.[66]

---

[65]https://www.ncbi.nlm.nih.gov/pmc/articles/PMC4810786/

[66]https://news.harvard.edu/gazette/story/2012/02/fears-of-bioterrorism-or-an-accidental-release/

However, as we can read from the abstract, it is important to note how the Dutch scientists focused mostly on possible cures for this strain:

"The transmissible viruses were sensitive to the antiviral drug oseltamivir and reacted well with antisera raised against H5 influenza vaccine strains.".

Antisera is the serum containing antibodies, in this case for the H5 protein.

The eerie part is the team reached this result by swabbing the ferrets' noses with common avian H5N1, slightly edited through site-directed mutagenesis. They swabbed the first ferrets and infected other ferrets with the new "ferret" H5N1. This was done for a total of 10 passages. It was enough to allow H5N1 to become directly transmissible between ferrets, airborne. This means that H5N1 has the potential to do so with humans too. So far, all human cases were promptly identified and isolated on the spot, but imagine an outbreak in a big poultry farm with low hygiene, where all the humans working there get infected and start exchanging the virus with each other...?[67]

Something similar happened 10 years later, with wild foxes... For the first time, ever, H5N1 was also found in wild mammals. The area is still the Netherlands, the animals are red foxes, two poor 6 weeks old cubs, found sick by the wildlife rescue service, in May 2021.[68]

"May 10, a red fox Vulpes vulpes cub (cub 1) displaying abnormal behavior was found in Bellingwolde, the Netherlands, and taken into care of a wildlife rescue center. Upon entry, the 6- to 8-week-old cub was slightly dehydrated and showed at intervals of <30 minutes lip retraction, rapid opening and closing of mouth, excessive salivation, skin twitching, head shaking, and body tremors."

---

[67]https://web.archive.org/web/20120630154419/http://news.sciencemag.org/scienceinsider/2012/03/breaking-news-nsabb-reverses-pos.html
[68]https://wwwnc.cdc.gov/eid/article/27/11/pdfs/21-1281.pdf

This poor little fox was then euthanized as his conditions were worsening fast.

"On May 13, the center received another 6 to 8-week-old red fox cub (cub 2) found ≈900 m from cub 1. Cub 2 was hypothermic and dehydrated. It had seizures and died overnight."

Even the presumed mother was found, in the same area, displaying the same symptoms:

"Retrospectively, we concluded that the mother of the cubs was likely a vixen found walking circles on May 10, ≈975 m direct distance from cub 1 and ≈90 m from cub 2. The vixen reacted very aggressively to capture, responding to sound but blind. The vixen had a fresh elbow fracture, probably caused by a road traffic accident. We humanely euthanized her the same day and sent her carcass for destruction."

Given the symptoms, the first virus they tested for was rabies lyssavirus, through antibody test on brain's tissue. The test turned negative. Only after this, they tested for H5N1, through PCR:

"Subsequently, we tested brain samples for avian influenza virus by using a PCR detecting the influenza A virus matrix gene, followed by the subtype-specific H5-PCR on the hemagglutinin gene, as described previously (2). The samples from both cubs tested positive (Table), and we subtyped the virus as highly pathogenic avian influenza (HPAI) influenza virus A subtype H5N1."

And then someone still says the PCR test is useless...

What scares me most is that the Emperor managed to reach wild mammals faster than I'd have ever expected. Moreover, he managed to infect foxes, genetically so close to our best friends. If H5N1 is able to proliferate among foxes, how long will it take for the virus to jump on dogs or cats? Before reaching humans, I'm afraid the Emperor will inflict much emotional pain on us, by taking away our beloved friends. Even before starting, the avian influenza pandemic will make the coronavirus pandemic look like a funny exercise.

And then, what will happen when H5N1 will become able to pass from human to human? I'm not an alarmist, but in this case, there are all the ingredients to be. Forgive me if I will go wild with my imagination, but for once I want to play the doomsday prophet role. You are free to believe or dismiss the scenario I'm going to predict. Call it a mental exercise.

The day the Emperor came

Pain and despair will hit the world even before the Emperor will conquer humans...

It's been only a few years since H5N1 was found in wild foxes, in the Netherlands. Now, the virus is found even in cats and dogs. Our pets begin to die overnight. The mourning owners face huge emotional pain, but this won't spare them from additional trauma. People are already aware the H5N1 virus has the potential to infect humans. There were already a few human cases here and there. Who lost their pet is isolated by society. Even those who have healthy pets, are seen with suspicion.

It takes only a few months to consider cats and dogs, "angels of death". Because of this, the first violent clashes take place. Some authorities consider the possibility of "getting rid of pets". Some authoritarian countries ban the possession of pets and proceed with atrocious pet genocides. They underestimate how pet owners can love their animals. To defend their friends, owners set entire cities on fire. The anti-vax protests seen during the coronavirus pandemic look like carnival parades now. The public opinion is still in favor of pet owners, but it's slowly shifting to a more drastic approach. The opposers, who fear the virus, call animal lovers "insane". Human rights are directly opposed to animal rights. The favor shifts considerably to animal-hate, once cats and dogs begin to infect their owners on a daily basis.

Media fuel hate for domestic animals just the way Nazi propaganda fueled anti-Semitism. Half of pet owners start to die. The witch hunt begins. Common people kill every animal they see and more and more often, they kill even the owners. Authorities start to turn their gaze the other way when a human friend of animals is killed too. A necessary sacrifice to maintain public health. Who still loves animals is considered irresponsible. Animal lovers respond accusing the authorities of not having prepared a vaccine for cats and dogs to stop the virus beforehand. Now, many believe it's late and that the virus will soon be able to jump from human to human.

Mass vaccination campaigns on humans start, thanks to the proto-vaccine available. The vaccine is made mandatory in the outbreak areas. The fear is so strong that very few oppose the vaccine. Animals are finally vaccinated too. After one year of chaos, the situation seems under control again, but society is now deeply divided. Homicides are skyrocketing. Pet owners are seeking revenge against who disposed of their animals. Also who lost their relatives due to contact with cats and dogs, are bound to kill pet owners. People start to miss when their nations were simply divided into left or right. Those days were much easier!

The hardline animal lovers do whatever they can to save the creatures from persecution. In the most civil countries, authorities and volunteers work to create biohazard proof sanctuaries, where to host animals safely, isolated from the human world.

After a short period of apparent rest, somehow, hospitals begin to be assaulted with avian influenza cases. However, this time, most people don't have pets. Only now, we realize the inevitable happened: H5N1 has become able to spread from human to human. Lockdowns are called in, while new RNA vaccines are promptly developed to tackle the new H5N1 human variant. This provokes unrest, fueled by a new season of disinformation and conspiracy theories. Ex-pet owners are still fed by revenge and decide to defy lockdowns. In response, authorities are way more coercive than they were during the COVID-19 pandemic, given the much higher mortality rate. They know they cannot afford to let this new virus go free. Too many cases would need hospitalization, so everyone would die. Authorities decide to use extreme methods. Even in the West, governments invoke any possible emergency state, any legal loophole or even constitutional amendments to use the force against the transgressors. Fines are just the first warning. The second warning is the baton. In underdeveloped countries, governments prefer the old, but still effective method of mass executions, supported by mass graves.

The internet is stormed with appeals to fight back for freedom. Clashes between authorities and the population intensify and get more violent each day. The police starts to kill citizens and citizens start to kill policemen. The secret services infiltrate illegal gatherings and protesting organizations. The risk of having agents among friends increases paranoia. Couples and families are broken apart by this paranoia. Domestic violence reaches all-time highs, due to a combination of this paranoia and never-ending lockdowns.

The economy is dead. Governments are forced to print fresh money and directly credit their citizens on their bank accounts. Who can afford it, puts all the savings into Bitcoin, which tops above $1 million. The stock market, on the other hand, drops like never seen before. Every attempt to put liquidity in the market is vain. Companies begin to be delisted to avoid further damage, including some Fortune 500 firms. Unemployment reaches record levels and so suicides. Food rationing is becoming the norm in more and more countries. As most people are incapacitated to work, thus to buy food, state controlled food supply seems the only card to play. West powers are forced to take inspiration from North Korea, the ultimate humiliation for the free world. These measures spark anti-communist conspiracy theories which become the new internet trend. From material for alternative nerdy media, these theories become mainstream media domain and end up in the official discourses of the political opposition.

However, going out to work means high risk to die. Every time a worker from the food industry dies, it's a huge loss for society. Alongside doctors and nurses, the heroes of the bird flu pandemic are the farmers, butchers, fishermen and everyone involved in producing and distributing food to the citizens in need.

At this point, some developed countries play the Swedish role. They give up on harsh restrictions such as the lockdown. Their reason is that there is no meaning in living in terror like that. We can't think to protect people if we start to arrest them or even kill them. This sparks outrage in the other powers. If country X lets the virus run free, it will soon develop into a more contagious and deadly variant, even capable of escaping the vaccines. Here are two possibilities:

1. Country X doesn't give up. War is declared on it by all the UN powers. Country X is brought down to its knees and forced to keep its citizens locked.
2. Country X gives up. The pandemic remains under control, just for a bit longer.

Let's go for number 2, which is more likely. Diplomacy can be so effective sometimes...

Due to increasing population suffering, measures are slightly loosen up. After 2 years into the pandemic, lockdowns remain only for the over 65 and people at serious risk. Restaurants are open again, with all the cautions possible. However, people start dying again. The virus escaped the vaccines. Vaccine makers are fast in editing their RNA vaccines, but the distribution chain is too slow to update the new product. After one year of constant contact with all the new variants, doctors and nurses start to get sick and half of them die. Hospitals become less and less capable of offering aid. People start to die of every disease and injury as nobody can assist them properly. Stress among the medical workforce causes more and more mistakes. Who didn't understand that the apocalypse has come, is thinking to make money by suing doctors for their human mistakes. Some surviving doctors start taking their own lives, smashed by a sense of guilt.

Meanwhile, in some parts of the world, some understand the apocalypse has come and start to throw mass illegal parties. Orgies are more and more common. The Emperor is aroused too. Now he can spread from body to body so easily. He loves this promiscuity! In just 28 months since he became a human virus, Emperor H5N1 killed 30% of the human population directly and caused the indirect demise of another 20%. Because of his nature, so different from President Corona, the Emperor took the life of almost everyone over 65, but also of most infants and children. He didn't spare the young adults either, but their stronger immune system allowed them to increase the survival rate. The good news is that most of the old people are gone. At least now, there are not many pensions left to pay and less patients in need of assistance for common diseases. We have not many doctors left after all this mess... The bad news is that most of the children are gone too.

If the virus could talk to the average adult now, he would say:

"I will come from where you expect me the least: from your very furry friends. I will kill them and then, I will force you to kill them too. However, when you will understand how vain your sacrifice was, it will be too late. I will be already behind your back. But I won't stab you.... I will stab your parents and your grandparents. After I'll have taken them, I will choke your children. They are so cute... And when you'll think I've taken all, only then I will flip my coin and kill your love too. I will let you live, so you can fall into complete madness. From then on, you will be a desperate beast, ravaging a post-civil world..."

After the apocalypse

A post-H5N1 world will look like Mad Max, just much greener. For Mother Nature, this era will be a real Renaissance. For the surviving humans, it will be an era of hardship. The beginning will be like cycling uphill. Humans will have lost many qualified professionals and many old individuals with their precious experience. The world will likely be made by the strongest individuals, but also by the most cynical, by those who have lost everything and everyone. It won't be a world for snowflakes. Words like kindness, empathy, respect, or inclusion, so important nowadays, will be set aside.

It's very possible that the culture emerging by such a traumatic event will be aggressive, a collective consciousness seeking revenge, looking for someone to blame for this chaos. Aggressive individuals will take many power posts and this will incite violence in the population. Society may well have strong macho traits and so, women might be relegated again to class 2 citizens, if values like aggressiveness, physical strength, domination and other typical masculine traits will be considered essential qualities to lead a community. A community which will be called to rebuild civilization. Imagine post-Soviet Russia.

Sure, there will still be space for nerds too. Those highly qualified individuals will be as precious as one Bitcoin. Cryptocurrencies will be used to pay these precious professionals. Those who didn't possess particular skills in the old world, will fill the void the way they can. For example, many doctors and nurses will be self-made. Their skills won't be exceptional, but will be enough to restore a decent health care system. In the areas most hit by the virus, crimes will be much higher than in the past, as there will be less police staff to enforce the law.

Humanity will have also lost most of its future. The Emperor didn't leave many children behind. This will be the most shocking difference with President Corona. The President never hurt children. Unfortunately, the Emperor is not Corona. The Emperor is a real piece of shit!

Nevertheless, the steady decrease in the world population will relieve many major problems we face today. The resources will be split among less people and thus, it will prevent new wars and it might even help to cease the current ones. Such an apocalypse will be a unique opportunity for humanity to start off again.

Yes, the Emperor will be a radical-change bringer, but the price for this change will be enormous. I still think that most of you prefer President Corona's soft line.

What I have described above is what would have happened if we'd have faced a bird flu pandemic without going through the COVID-19 pandemic first. It would have been an apocalyptic disaster. Even the notorious Disease X simulation, run in May 2018, confirmed the unpreparedness of the authorities against a possible global pandemic. COVID-19 has been just a Disease X-style exercise, run on a global scale, using a real, yet mild virus and by involving every individual of our society. President Corona was the coach to train us for something we had no idea how to respond to. Thanks to this massive exercise which COVID-19 was, now we are able to stand properly against an influenza pandemic, the real threat, the real enemy. If we use our training properly, I'm sure that the first human-to-human H5N1 cases will be promptly isolated and the disease will never spread among us or neither among our furry friends! Thank You, President Corona!

Opportunities Offered by President Corona

Look around and try hard to see what good opportunities this pandemic brought. If you want to be successful in life, you need to learn how to see the bright side in the darkest hour. I know, it's hard and you will even feel stupid at the beginning, but developing this skill will give you astonishing results. This took me out of the worst moments and led me to an ever-improving life!

Let's be creative and even mean. Let's go wild with imagination! Let's explore the opportunities offered by President Corona!

Hiding Your Dirty Business

If you are a politician, a manager or any person with responsibilities, a pandemic is a great opportunity to do things people don't like or don't want!

Whether you want to:

- Genuinely help your country but you'll need unpopular measures,
- Implement authoritarian laws and erode your citizens' rights,
- Or simply plan to use your own position for your personal gains (#corruption), a pandemic is the best moment!

The very Chinese government gave us a great lesson of pragmatism. They turned a problem into a massive opportunity! China is running concentration camps (Yes! Though government officials will call them "re-education camps") in Xinjiang (North of the country) to brainwash the local Turkic minority of the Uyghurs. The pandemic allowed China to carry this barbaric operation far from the international media's lights. The pandemic offered a great aircover from this undesired attention.

Not only governments though. If you run a criminal organization or your business is still not an example of integrity, the chaos sparking from the pandemic was a unique chance to make that huge transfer of funds you've always been eager to do. Maybe it was time to renew your henchmen's arsenal, but it's not easy to make the payment for hundreds of guns for private use go unnoticed. You need to launder that payment first. Thanks to the pandemic, many AML measures were relaxed. Banks wanted to make payment processing faster, especially when those payments were bound for aid, medical equipment, or charity. Never like during Covid, charity was so needed to help those affected by the sudden restrictions. Criminals could use these cracks in the security nets of the financial system to camouflage their dirty business operations.

Charity is not the only channel for laundering your dirty money. Given the explosion in 2020, online gambling is a perfect oasis too. With people locked at home, the online gambling customer base increased dramatically. Many companies found themselves understaffed, making the usual checks more difficult. At PokerStars we handled the situation impeccably, but I'm aware that most smaller companies struggled with the new volume. It's obvious that many suspicious activities could not be checked and went under the radar. Money-laundry professionals knew how to profit from the situation and get camouflaged with the rest of the exploding customer base, laundering billions during 2020.

What are the most common activities the money to launder comes from? You can guess it:

- Arms trafficking
- Drugs trafficking
- Cyber-crime, like hacking and identity theft
- Human trafficking

I don't approve of the war on drugs. It was proved that treating drug addicts as criminals is ineffective, besides being inhumane. If drugs should not be a problem, preventing money laundering is fundamental to fight horrible crimes, above all human trafficking. Money laundering is not a victimless crime. During my career in online gambling, I was shocked to learn how, in our times, human trafficking is on a steady rise, becoming one of the most profitable sources of income for criminal organizations!

According to the International Labor Organization, as of 2016[69]:

- *An estimated 40.3 million people are in modern slavery, including 24.9 million in forced labour and 15.4 million in forced marriage.*
- *It means there are 5.4 victims of modern slavery for every 1,000 people in the world.*
- *1 in 4 victims of modern slavery are children.*
- *Out of the 24.9 million people trapped in forced labour, 16 million people are exploited in the private sector*

[69]https://www.ilo.org/global/topics/forced-labour/lang--en/index.htm

such as domestic work, construction or agriculture; 4.8 million persons in forced sexual exploitation, and 4 million persons in forced labour imposed by state authorities.

- *Women and girls are disproportionately affected by forced labour, accounting for 99% of victims in the commercial sex industry, and 58% in other sectors*

Here we are talking about real slavery, not about your local mayor imposing to wear a face mask, you spoiled, selfish, illiterate snowflake! Do you know how much all of this shame is worth? It is estimated that the global forced labor market is worth more than $150 billion. To give you an idea, consider the drug trafficking market is estimated to be between $426 billion and $652 billion. Drugs are worth just 3-4 times more and honestly, sniffing a line is not comparable to enslaving a human being. Better a million of cocaine addicts than a single pimp.

Maybe, instead of hiding, you want to set up a dirty business out of the pandemic. Many did so, from embezzling funds bound to fight COVID-19, to scamming authorities with false promises of medical equipment delivery.

In Brazil, between April and November 2020, the Federal Police arrested hundreds of individuals for corruption or misapplication of public resources and money laundering, in 17 different states. The money came from public tenders and public contract frauds, which together amounted to BRL 1.9 billion (approx. $360 million). The crimes ranged from the purchase of unlicensed or fake medical equipment, to corrupting officials for winning public tenders. Something very old-fashioned became fresh again during the pandemic.

In March 2020, German authorities were conned by a Nigerian scheme. A dummy company set in Spain promised the Germans the delivery of 1.5 million face masks, for which an up-front payment of €1.5 million was required to an Irish bank account. The problem is that no company was in Spain. Their website was fake. Unaware, the German authorities were referred to a dealer in Ireland, turning the Germans again to another vendor in the Netherlands, which allegedly had the masks. At that point, they said an additional payment of €800,000 was requested. This triggered the alert, which even called in the Interpol to investigate. The banks involved managed to freeze the transaction before it could reach a British bank account. From this account, the funds were bound to another account in Nigeria.

Such a gig for that money. A gig which was terminated in the UK, right at the climax... Damn it!

These are two shocking examples, but cases like these are countless. If we talked about all the fake sanitizers, fake masks, even fake ventilators "sold", I should write another book!

Nothing like a crisis offers new opportunities for con-artists. Keep your antennas vigilant during these times. COVID-19 can really be a scam, if you approach the wrong people...[70]

If instead if you plan to be the mastermind of these schemes, well go for it. But keep in mind that whatever you do, soon will hit you back, with the interest.

Possible China's Collapse

The collapse of China sounds tempting for many lovers of human (and animal!) rights. It's even the forbidden American dream...

With this pandemic, China has lost much of its credibility and trust. Businesses could consider relocating to other countries that by now are much cheaper than China itself. This could be a great occasion for other Asian countries like the Philippines or Indonesia, but even Africa! Africa is now ready to receive investments rather than charity, something that China itself understood long ago. Chinese entrepreneurs could be the authors of their own country's demise if they start producing

[70]https://www.fatf-gafi.org/media/fatf/documents/Update-COVID-19-Related-Money-Laundering-and-Terrorist-Financing-Risks.pdf

in Africa. It's important to notice, indeed, that most of the massive Chinese investments in Africa derive from private Chinese companies, not from the Chinese government itself, unlike we tend to believe in the West. Chinese companies are even changing their attitudes. They no longer "go to Africa", but rather they try to "set roots in Africa". That's why their commitment in the continent increased during 2020 by a neat 8%, in spite of the pandemic.[71]

Let's see other reasons for which China's collapse would be a good present from the virus:

- Other developing countries would benefit from investments and new manufacturing businesses,
- Western countries could regain some specific manufacturing industries,
- Western countries should develop the new 5G infrastructures on their own for the happiness of the security experts,
- Such a crisis would stop China's Belt and Road Initiative, which de facto aims to have global business (and then politics) under China's control by 2049,
- China would no longer be taken as a model, especially on politics as too many thinkers even in the West deem the Chinese system "superior" due to lack of democracy, freedom and human rights, seen as "obstacles". The "free world" would win once again over the totalitarian world!

However, don't think this is so simple! A collapsing China could even have dire consequences.

First, China is not as homogeneous as you think. Like every huge country, it has many ethnic groups and cultures, resilient and resistant even to the harsh communist government. So resistant that the government is being forced to implement corrective camps, the ones we talked about before. So:

---

[71] https://furtherafrica.com/2021/09/02/report-the-main-destinations-for-chinese-investment-in-africa/

- A collapsing China would be the right occasion to unleash separatist movements across the country, from Hong Kong to Xinjiang,
- Mongolia could seize the moment to claim the lands of Inner Mongolia,
- These conflicts would inevitably attract the intervention of foreign superpowers, rising the risk of a new global conflict, i.e. WWIII,
- To beat the competition of the new manufacturing countries, China might reverse on worker rights where the nation improved dramatically in the last decades, reintroducing for example minors or forced labor,
- The government could reverse on some freedoms and make China a giant North Korea very hostile to the outside world,
- A China in crisis could starve, starting to eat dogs again, ruining all the progresses done on animal rights.

It's probably easier to calm China down and let the CCP die of natural death in the long term. It will happen sooner or later. We just don't know when...

Reshaping the European Union

This crisis showed how the European Union, but even the single European states were unprepared to face such an emergency in an effective way.

I'm sure many of you have said at least once: "The EU is an unelected body!"

Can you remember where you heard this, the first time? Who told you that? Most probably you heard it from Nigel Farage, during his Brexit campaign. This is one of the several lies Mr Farage said about the EU. In the EU, you elect at least two bodies: one directly, one indirectly. In one of these two, you sent Mr Farage himself! With your own very vote!

What the EU Is and How It Works

Every 5 years, you vote in the so-called "European elections". This event has happened since 1979 (in those countries which had been EU members since then). These elections allow you to choose the members of the European Parliament (MEPs). The other legislative chamber of the EU is the Council of the European Union, composed by the Ministers of each EU member. This organ is indirectly elected, as it's formed by the people you have already chosen in your national elections.

Beside the two chambers, you have the European Commission, a sort of European government, and the European Council, which acts as a collective president, formed by the prime ministers of each EU state.

The European Parliament counts 705 members. They are proportioned to each state's population. The main tasks of the European Parliament are to:

- approve or amend legislative proposals submitted by the European Commission,
- approve or reject the European Commission members chosen by the European Council,
- approve the EU budget,
- approve the European Central Bank's president,
- supervise the other EU institutions,
- initiate inquiries and other legal actions against member states which breach EU laws.

It's important to notice that the EP pursues the general interests of the EU. MEPs work on common legislation, they don't work to defend the particular interests of their home-state. Not in this assembly.

If the MEPs have a law idea, they can send the draft to the European Commission, which in turn will work on the draft and present it before the whole parliament, as it happens in the national legislatures. These laws are called European Directives. These are direct expressions of European law.

Member states have a time-frame to implement them in their national laws. This is to harmonize legislation between member states, especially when it comes to commerce, technology and finance. It's thanks to these directives that you can buy a car in Germany without any further homologation, even if you live in Belgium.

The EP is a very important organ which has sensible repercussions on our everyday lives. All the European Directives you hear about, pass through this chamber. It's important to understand its primary functions and to be aware of what we are doing when we vote at the European elections, every 5 years.

The other legislative organ, as we said, is the Council of the European Union. This council includes 27 seats, one per member state. The presidency of this council changes every 6 months, from state to state. That's why you hear: "This year Poland has the presidency of the European Union!". It means that Poland is presiding over the Council of the European Union. The 27 ministers sitting at this table change based on the topic discussed. If the topic is public health, the health ministers of each country will gather. If the topic regards agricultural affairs, the ministers of agriculture will sit here. These representatives can pursue and defend the interests of their home-states. It's here that we saw a lot of epic clashes during the early stages of the COVID-19 pandemic.

The CoEU functions are similar to the ones assigned to the EP, but the CoEU brings in a specialistic cut to the bills discussed by the EP and the EC. The CoEU is also entrusted with the task of choosing the members of the European Commission, which will need to be approved by the European Parliament.

When it comes to highly sensitive topics, like the approval of the extraordinary economic measures conceived during the COVID-19 pandemic, the CoEU usually requires unanimity to approve a bill or a resolution. That gives the member states a power of veto.

When instead the CoEU discusses ordinary matters, like approving or amending a law coming from the EP, the approval requires a simple majority of 55% of the members, representing at least 65% of the European population.

The other primary body is the European Commission, a body which resembles a national government at a first look, but differs from it in many ways. Even the European Commission is composed of 27 members, one per each state, but these members have to pursue the general interests of the union, rather than the ones of their home-state. Like a national executive, the EC has to:

- manage and direct day-to-day operations,
- propose new EU directives and laws,
- enforce EU laws.

In layman's terms, the EC reviews every new law approved by a member at national level. If this law does not comply with the EU law, the EC can suggest to the national parliament how to amend the text and make it compliant with the EU law. If the member state refuses to do so, the EC can sue the member state to the Court of Justice of the European Union. According to the outcome, the EC can issue fines and other penalties as expressed in the EU treaties.

Yes, you're right, the European Commission is not directly elected, but very indirectly. The candidates are proposed by the European Council and approved by the European Parliament, if they like them. If the EP rejects those candidates, the Council must work on a new list to present. Over the years, there have been many talks and proposals to get the European Commission elected directly by the European citizens, alongside the elections for the European Parliament. Many so-called "Euro technocrats" are actually favorable to the idea. Who's against, might surprise you, are the very national governments, even those that at home call out the lack of democracy in the EU, but once in Brussels, are the first ones keeping the European population from being involved more. If you believe that voting directly for the European Commission is a fundamental right of the European citizens and a fundamental step to make the EU really democratic, you may consider starting a petition on the coolest institutional portal ever, made available by the very EU. This portal bears the cute name "Petition the EU".

Any individual or organisation based in the EU can submit a petition to the EU on an issue related to EU policy or law. Find out how to create and submit one.

https://ec.europa.eu/info/about-european-commission/get-involved/petition-eu_en

This can be a very powerful tool that most national governments don't even have. Not bad, for an institution that is always accused of being anti-democratic!

Between these three bodies, there is a "liquid" institution, an intergovernmental body called the European Council. Media used to call it "EU Summit". It's the assembly of all the prime ministers of each member state. The European Council has no legislative power, but it has a great influence in determining the EU agenda and pushing the European integration process between the member states. It is mostly a body useful to discuss and tackle emergency situations like the Sovereign Debt Crisis in 2011 (also known as Eurozone-crisis) or the very COVID-19 pandemic. We still remember the battle between the North states and the Southern states over how to fund a recovery plan to help the pandemic hardest-hit countries.

I define the European Council "liquid", because the Council's role can change according to the situation it has to face. Given its nature, this Council summons on average 2-3 times a year, depending on the needs.

Beside its fireman role, the European Council has few, but crucial tasks within the EU:

- Propose the possible European Commission's members and submit them to the European Parliament for voting,
- Propose candidates for the European Central Bank

presidency and submit them the European Parliament for voting,

- Aid the EP and the EC in negotiating treaty changes,
- Aid the EP and the EC in negotiating international treaties with third nations,
- Define the EU foreign policy and assure the political line remains uniform among all members.

The latter task became more and more prominent in the last years, especially after the involvement of European powers in the Middle East conflicts. To better tackle this task, the Council created the High Representative of the Union for Foreign Affairs and Security Policy. I know, in Brussels they are fond of complex, long or ambiguous names. This is one of the things I hate the most about the EU. You're right about this. This pompous name became important in defining not only a common EU foreign policy, but also a common EU security policy and since the UK's departure, this office became a powerful promoter of military integration between the EU states.

Of course, keeping a unique political line among 27 members is not easy. 27 nations might have very different needs and interests in foreign policy. These needs may derive by their economies but also by historical relationships.

The very European integration process is something bound to take decades to complete. In Brussels, they usually try to see the gains you can get in years (look at the Green New Deal goals, set for 2050!), while national governments rarely look beyond their mandate. They don't, because their own voters are not used to looking so much into the future. Voters are impetus and emotional, ready to change opinion based on a headline they have read while sitting at the loo.

When voters are not happy with what their national governments, or even the EU itself, are doing, they can appeal to the Court of Justice of the European Union. This is a judicial body independent from the other EU institutions. Its role is assessing if what the EU and the national governments do and decide is compliant with the EU Law, as defined by the EU treaties. It is the most powerful juridical tool that citizens can use against the abuses and wrongdoings perpetrated by political authorities, both at European and local level.

The other watchdog body within the EU is called the European Court of Auditors. This independent body checks how the EU spends the taxpayers' money. Thanks to these guys, we can know where our money goes once it is sent to Brussels. Yes, the EU is not so shady, but a big merit goes to this court. If this body didn't exist, I don't know what would happen to our money. Yes, of course there will always be something slipping before the Court's nose, but still crumbs compared to what would happen if no-one was watching.

Europe Needed a Common Enemy

Lack of foresight led many European governments to cut on the healthcare system, in order to comply with the stringent Stability and Growth Pact (SGP) outlined by the EU institutions. This pact, deriving from the Maastricht Treaty, imposes members of the EuroZone, to keep their public debt under control. The government deficit shall not exceed 3% of GDP and the national debt should stay lower than 60% of GDP. To comply with this, over the years, many governments took the austerity shortcut, cutting mostly the public services, including the healthcare system.

Now, President Corona showed that cutting expenditure on public healthcare was not a good idea. It showed that Europe is far from being united, not only at continental level, but even at local level. Local divisions within the EU states came out clearly during those chaotic days.

Don't get me wrong, I'm a pro-EU and even a federalist. I'm perfectly aware that 90% of the things said about the EU are bullshits. I prefer to build rather than destroy. Destruction is for losers, indeed most anti-EU fans are losers in life.

Nevertheless, I truly believe that something must change, not only in the EU institutions, but in the very relations between member states. Solidarity is something basic that must be ever present in the continent, with or without the European Union. There is no excuse for how every party involved pursued its own interests or worse, tried to profit from the situation at the expense of the weakest ones!

When we will learn to be one, then we will be able to work on common institutions. But if not even a common enemy will unite us, I doubt something else can do it...

Maybe the threat of Russia looming over Eastern Europe might unite the continent. Suddenly, a human enemy is more motivating than a virus enemy. It might encourage European military forces to cooperate closely. If there is something where the EU can really make sense, it is to build a common European Army. An army of this kind would be bigger, yet more efficient and less expensive for the taxpayers, than national armies. Curiously, even in this season of Euroscepticism, Europeans still want an European Army. According to an Eurobarometer survey, 55% of Europeans wish for an EU Army, while 75% confide at least in a common defense and security policy for the continent. The favour is strong even in those states led by Eurosceptic forces like Poland (68%) or Hungary (75%).[72]

Few know that, since the UK left the EU, the bloc made big progress in this direction. The UK had always been the great opposer to a European defense cooperation, so no surprise that the most important measures were taken after the 2016 Brexit referendum. Since 2017, the EU has begun to implement ambitious initiatives to provide more resources, stimulate efficiency, facilitate cooperation and support the development of capabilities, as reported by the European Parliament:

---

[72]https://www.europarl.europa.eu/news/en/headlines/security/20190612STO54310/eu-army-myth-what-is-europe-really-doing-to-boost-defence

- *Permanent structured cooperation (PESCO) was launched in December 2017, and 25 EU countries are participating as of June 2019. It currently operates on the basis of 47 collaborative projects with binding commitments including a European Medical Command, Maritime Surveillance System, mutual assistance for cyber-security and rapid response teams, and a Joint EU intelligence school.*
- *The European Defence Fund (EDF) was launched in June 2017. It is the first time the EU budget is used to co-fund defence cooperation. On 29 April 2021, MEPs agreed to fund the flagship instrument with a budget of €7.9 billion as part of the EU's long-term budget (2021-2027). The fund will complement national investments and provide both practical and financial incentives for collaborative research, joint development and acquisition of defence equipment and technology.*
- *The EU strengthened cooperation with Nato on 74 projects across seven areas including cybersecurity, joint exercises and counter-terrorism.*
- *A plan to facilitate military mobility within and across the EU to make it possible for military personnel and equipment to act faster in response to crises.*
- *Making the financing of civilian and military missions and operations more effective. The EU currently has 17 such missions on three continents, with a wide range of mandates and deploying more than 6,000 civilian and military personnel.*
- *Since June 2017 there is a new command and control structure (MPCC) to improve the EU's crisis management.* [73]

Before establishing an EU Army though, we need to calm the tensions that still exist among the European nations.

---

[73]https://www.europarl.europa.eu/news/en/headlines/security/20190612STO54310/eu-army-myth-what-is-europe-really-doing-to-boost-defence

Clashes During the EU Summits

The weeks that followed the first COVID-19 outbreaks in Europe set the meeting rooms in Brussel on fire. The very destiny of the Union was at stake.

Hard times make strong men. While the press was highlighting the apparent feuds between European states, particularly North versus South, I was delighted to see such intense debates over many aspects of the Union that were left aside for too long. From established politicians, going through European and local institutions and ending with private citizens, I saw a level of involvement and proposals never met before about the shaping of the EU! It's like if President Corona gave Europeans an ultimatum: what do you want to do with your Union? Do you want to unite properly or do you want to go back to your nation states? Do you want to have a prominent role in the world or do you want to take care of your garden and let the other superpowers decide the destiny of the planet?

These questions were already posed in the Brexit aftermath, but on that occasion those questions were almost left unheard by Europeans. In 2020, the President reformulated those questions in a more assertive way. After all, it's not important what you ask, but how you ask for it. This was the final showdown. If a global emergency doesn't bring Europeans together, nothing else will.

As of writing, it seems that a remarkable accomplishment of President Corona is actually to speed up European integration. The virus forced European leaders to face issues left behind for too long, like financial mutual assistance, a possible common debt to fund EU actions, functional European-level Civil Protection and CDC with real decisional power, to act fast when and where needed and most importantly, the future of the monetary policy. Finally, the strict Stability and Growth Pact was suspended and now, European states don't have to worry too much about their debts.

The SGP was conceived by German Minister of Finance, Theo Weigel, in the 1990s. Germany had based its strong economic performance on a low-inflation strategy, which contributed to keep prices stable and wages high. Weigel didn't take long to convince his European colleagues that extending this approach to all the other states was a great idea. It would have helped them to lower their inflation, which had been astronomical during the 1980s. However, to keep compliant with the SGP, a state had to ensure a rigid fiscal policy, meaning that the tax evasion had to be very low, if a government didn't want to take the way of austerity. This soon became a problem, especially for the Southern states, where tax evasion was traditionally higher than in the North, while citizens were very reliant on public services and subsidies. Being incapable of tackling tax evasion properly, some states like Spain, Italy and Greece, cut progressively their public expenses, even on vital areas like education and healthcare. The global financial crisis in 2008, exacerbated the austerity policies, until the Euro-zone almost collapsed when Greece became incapable of keeping within the SGP. This financial crisis provoked the first serious anti-EU wave. In a few years, Europeans became disenchanted with the European Union project. Until the 1990s, most citizens of the continent were enthusiastic about an economic and political integration with their neighbors. Referenda to join the European Union registered peaks of 70-80% of consent, like in Ireland in 1973 or Czech Republic in 2003. In the wake of the Euro-crisis in 2011, surveys recorded that enthusiasm dropped dramatically among Europeans. The dream of turning Europe into a proper federation got cold when the common currency entered into play. People were missing the apparent economic freedom their nation could enjoy before entering the Euro. Unlike Weigel foresaw, citizens felt like prices increased while wages stayed the same. Many economists argue that creating the Euro was a good step in principle, but that the timing was premature. Moreover, ministers took for granted that what worked for Germany, would have worked for the rest of Europe as well. Some saw the Euro as the continuation of the German Mark, an aggressive act of Pan-Germanism, a new attempt from Germany to control the continent, no longer through military warfare, but through financial warfare. Jumping to this conclusion when your own national economy suffers is a human reaction, but that doesn't mean it makes sense. If we look at the situation from above, it is clear that, if the European economies go downward, Germany's economy will suffer too. In Germany's view, the SPG was a tool to help the other members to flourish, so that Germany could keep growing too. The problem with the SPG has always been its lack of flexibility. This aspect has always been criticized by many leaders, of any political area. When your debt can't exceed 3% of your GDP, not even a single year, everything becomes difficult. The SPG would make more sense if it takes into consideration a full economic cycle, which usually takes 5-6 years.

Thinking out an exit strategy was everything but easy for the Union. President Corona wanted to highlight the deep cultural divide cutting the Old Continent in two: the Southern states and the Nordic states. The President hit the South first, to remark on the bad financial policies of the past, highlighting how the debt situations in countries like Italy and Spain were no longer sustainable. This posed the Nordic states a dilemma: was it legit (or even a duty) to aid the Southern states and their still economies or ignore the SOS and let the Mediterranean pay for its past sins?

It's important to notice how the press in the two areas presented the problem so differently, too often failing to show the reasons of the counterpart, telling you only why either of your side was right and demonizing the other side and the EU as a whole.

In the South the problem was presented by depicting the Northern states as merciless, insensitive about the EU's future or even worse, trying to make the South collapse on purpose so the Northern economies could expand at the expense of the South. Some Southern Europeans were stressing that not helping the most hit countries was an insane assist to the Far Right, whose ultimate goal is the dissolution of the EU.

What the Southern press was ignoring was that the reverse approach (helping the South) was the very fuel for the Northern Far Right. In the North, the press often called out how much money was given to the South, money that according to them comes directly from the Northern taxpayers. In the Northern eyes, billions were taken from the "honest and hardworking people of the North" and given to the "corrupt and lazy people of the South". If this will happen again, it's time to dismantle the EU and move on our own...

Whenever you walk in the other's shoes, you can see the real complexity of an issue. It was clear that the pressure on the Northern leaders was as high as on the Southern leaders. This was primarily what made the European Councils in March 2020 so tense and tight. The first EU Summit in March 2020 is remembered for the shocking declarations of Dutch Minister of Economy Wopke Hoekstra, who asked the European Commission to start an investigation into Spain's finances to check if their request of aid was legit. Portugal's PM Antonio Costa labeled Hoekstra's declarations as "repugnant". But it's just because of such a vortex of discussions, proposals, brainstorming and mutual attacks too, that the most important measures in EU history were decided, approved and implemented, in record time. All of this thanks to President Corona... Let's have a look at what the President created for Europe!

The EBC Buys Everything!

Populists were asking their governments to print money. Japan adopted this strategy, the US printed a lot too, but looking at history, who prints a lot of money doesn't go far. Printing money means that initially, you can borrow this new money for free, but in the long term, that same money will lose value, raising inflation, the very thing everyone fears now. Instead of printing random money to make the populists happy, the European Central Bank created a new program to buy governments' debts, banks' bonds and other financial assets, at interest rates lower than the market. This means that the troubled economies had in the EBC a friendly buyer who bought their bonds, funded their debt, without expecting any significant interest back, which usually stands around 3%. This program is called the Pandemic emergency purchase programme (PEPP).[74]

"The PEPP allows the ECB to purchase different types of assets in financial markets. By doing this, the prices of those assets go up and, by extension, market interest rates go down. All of this supports the economy by making borrowing cheaper for people, businesses and governments. It also makes it easy for borrowers to service their debts. This helps to maintain borrowing, spending and investment despite the pandemic crisis. Ultimately, it will boost growth and bring inflation back to our target of 2% over the medium term, i.e. in line with price stability."

This package is worth €1,850 billion. In this way, the ECB provides liquidity to the European economy, but without skyrocketing inflation. Letting inflation go wild would mean that your salary will lose purchase power quickly. On the ECB website you can find a nice graph which explains how the system works. The fundamental steps are:

---

[74]https://www.ecb.europa.eu/mopo/implement/pepp/html/index.en.html

1. The European Central Bank buys bonds from banks.
2. This increases the price of these bonds and creates money in the banking system.
3. As a consequence, a wide range of interest rates fall and loans become cheaper.
4. Businesses and people are able to borrow more and spend less to repay their debts.
5. As a result, consumption and investment receive a boost.
6. Higher consumption and more investment support economic growth and job creation.
7. As prices rise, the ECB achieves an inflation rate of 2% over the medium term.[75]

ESM Without Commitment

The European Stability Mechanism (ESM) is an EU agency set to provide emergency loans directly to governments. It was born to incorporate under one umbrella all the bailout programs created after the 2008 financial crisis, when Greece and Cyprus fell, smashed by their excessive debts.

"The purpose of the ESM shall be to mobilise funding and provide stability support under strict conditionality, appropriate to the financial assistance instrument chosen, to the benefit of ESM Members which are experiencing, or are threatened by, severe financing problems, if indispensable to safeguard the financial stability of the euro area as a whole and of its Member States. For this purpose, the ESM shall be entitled to raise funds by issuing financial instruments or by entering into financial or other agreements or arrangements with ESM Members, financial institutions or other third parties."[76]

---

[75]https://www.ecb.europa.eu/explainers/show-me/html/app_infographic.en.html

[76]https://www.esm.europa.eu/sites/default/files/20150203_-_esm_treaty_-

President Corona made this instrument even easier than it was before. When it was established in 2012, the ESM granted emergency bailouts to the states in muddy waters, on the condition that the government pursued far-reaching reforms, which usually had a hard impact on public services, welfare and even the healthcare system. The ESM was designed almost as an extortion mechanism to push the weaker Euro countries to reform their economies. Asking a country to take painful reforms in order to access a loan seems harsh, but let's look at the issue from a different angle. Imagine you have a drug-addicted friend who asks you for some money. He's a friend you really care about. What will you do? Will you just give him a 100 bucks, knowing such money will end up through his nose, or will you demand that he join a rehab in order to get your money? If you really care about your friend and you want a permanent, positive change in his life, you will demand that he takes the necessary actions to stand on his feet again. The same applies to states drowning in debt because of their policies. Let's look at Greece, the very first client of the ESM. Greece reached an all-time low after years of high tax evasion, lavish public spending and countless frauds against the very European Union, which poured billions into Greece, billions which ended in the hands of the local mobs or to buy supercars.[77]

For sure, the treatment received by Greece has been brutal, but at least Greece is still in the international financial system and as of 2021, the Hellenic nation is projected to grow by 6%! Not bad, for a state that many declared bankrupt just 10 years before!

Now, thanks to President Corona, the ESM is being used to keep crippling economies afloat. The sole condition imposed by the ESM 2020 version is that the money must be used to directly tackle the coronavirus emergency and the loss of

---

_en_1.pdf

[77] https://www.dw.com/en/eu-funds-for-projects-that-never-existed/a-16113829

employment which followed. It also spared national governments from issuing new debt. Issuing new bonds during an unprecedented economic crisis would have meant that the single states would have to issue bonds with a too high interest rate, like it happened to Greece in 2009.

The wish is that states will never need the ESM again, but at least, should they need it, the money will be available with less daunting conditions.[78]

SURE

There was a lot of brainstorming during the EU summits held in March 2020, the first summits held completely as a video conference. Thank you, Zoom! The most hit countries in March 2020 (Italy, Spain and France) needed money as a COVID-19 patient needs oxygen. While the ESM mechanism was feared and opposed by nationalistic forces, a simpler and generous package was accepted to tackle the sudden unemployment which arose in the lockdown aftermath. This program was named SURE and it was the first solution the EU baked for the European torn economy.

"The temporary Support to mitigate Unemployment Risks in an Emergency (SURE) is available for Member States that need to mobilise significant financial means to fight the negative economic and social consequences of the coronavirus outbreak on their territory. It can provide financial assistance up to €100 billion in the form of loans from the EU to affected Member States to address sudden increases in public expenditure for the preservation of employment. SURE is a crucial element of the EU's comprehensive strategy to protect citizens and mitigate the severely negative socio-economic consequences of the coronavirus pandemic."[79]

---

[78]https://dobetter.esade.edu/en/coronavirus-pandemic-spain

[79]https://ec.europa.eu/info/business-economy-euro/economic-and-fiscal-policy-coordination/financial-assistance-eu/funding-mechanisms-and-facilities/sure_en

The SURE program revealed to be a prototype for a much bigger and more ambitious program, called NextGeneration EU...

NextGenerationEU

The main goal of the Southern states was to create the so-called Corona Bonds. These would have been bonds issued directly by the European Union, backed by the EU budget. In this way, all the 27 states would have shared a common debt. This was the main concern for the "Frugal Four": Austria, Denmark, the Netherlands and Sweden. These states are traditional supporters of a rigid fiscal policy and firm believers of the SPG, even more inflexible than Germany is. They usually don't trust the Southern states, as we could see in the flamed EU Summits of 2020. Germany often sides with the Frugal Four too, but this time around, Angela Merkel worked as an intermediary between the North and the South. As a true leader, Merkel tried to find the common ground among all the EU members. She understood that President Corona doesn't forgive childish skirmishes of this kind. It wasn't time to brag about who is good or bad at accounting. Merkel helped the EU members remind what the goal was: to exit from a dire situation and make the economy kick off again.[80]

The negotiations were tense. Eurosceptics were smelling the dissolution of the European Union. They began to believe seriously that the EU could be President Corona's next excellent victim. In the end, it turned out the President strengthened the Union. What doesn't kill you makes you stronger. NextGenerationEU is indeed the strongest economic action ever taken by the EU. It marks an historical victory for the Southern States. The game was won when President Corona started to ravage also the Frugal Four, during the

[80] https://ec.europa.eu/info/sites/default/files/about_the_european_commission/eu_budget/nextgenerationeu_green_bond_framework.pdf

second devastating wave. At that point, even the North realized that allowing the EU to issue bonds directly and share a common debt to fund a strong recovery was an excellent idea. Shit can happen to everyone, everywhere.

Everyone talks shit about the EU, but in reality, this institution enjoys high respect and credit among investors and rating agencies. This means that bonds issued by the EU will always rate triple A. When your bonds are triple A, it means that investors trust you so much that they will borrow you money for the lowest interest rates possible. This was impossible to achieve for the single states, after a year of lockdown and billions of GDP burnt like gasoline. The so-called Corona Bonds became the effective source of funding for the massive NextGenerationEU package, a recovery plan worth €750 billion, the biggest financial maneuver in EU history.

This program is way more articulated than SURE. The latter was a short-term bandage to tackle unemployment and lack of medical equipment in the immediate aftermath of the outbreak. NextGenerationEU is a multi-annual program that stresses on innovation. The main areas of interest are:

- The healthcare system, to make it stronger and better prepared for future pandemics, but also more capable of curing known diseases like cancer,
- Energy production, to make the EU totally green by 2050. This will not only grant clean cities for Europeans, but it will also mean energy independence from Russia!
- Digitalization of services. President Corona is wiping away paper bureaucracy with a speed never seen before with any other leader. The EU wants to rush headlong in this revolution and make the whole continent as digital as Estonia![81]

The plan is really ambitious and if it succeeds, we might look

[81] https://europa.eu/next-generation-eu/index_en

back at these days and say: "Those were the days the EU became a true nation!". I know you don't like this, but maybe your children and grand-children will. Think about them before fucking their future like the British grandpas did with their kids...

Now that you know how the EU works and what it has done over the COVID-19 period, you have all the tools to form a more comprehensive opinion about it. If you already liked the idea of a United Europe, you may have more reasons to believe in the project. If you are still Eurosceptic, that's great too. At least now you have a solid basis to attack the EU properly, without having to parrot Farage's empty slogans.

ECDC

Did you know that there's an EU agency called the European Centre for Disease Prevention and Control? It was created in 2004, to prevent the spread of infectious diseases across the Union. It was the logical response after the appearance of SARS in 2003.[82]

EU authorities wanted a body within Europe which could coordinate the response against a possible pandemic across the whole bloc of paramount importance. The main problem with the ECDC is that the EU members never wanted to entrust this agency with the budget and the power its role would demand. Before the COVID-19 pandemic, the ECDC budget was merely €50 million a year! Breadcrumbs compared to the $12 billion the American CDC enjoys each year. The American CDC is a clumsy bureaucratic mammoth, but at least it has the financial means to protect a 300 million people nation. The ECDC was incapable of preparing a prompt, quick response to COVID-19 when the virus appeared in Europe. Low budget means low means to prepare reports, make proper analysis and have the human resources to carry highly scientific tasks in a timely manner, and when it comes

---

[82]https://www.ecdc.europa.eu/en

to infectious diseases, time is everything!

To save money and time, the ECDC could have relied on data coming from the national health authorities, but unfortunately, the data sent by these has been inconsistent since the beginning. During the first half of 2020, each state was counting deaths and cases in its own way (in the very first days, some even tried to conceal cases!). As we said before, this is something that irritated even the WHO. Data was inconsistent also when it came to tests. Some states showed a much higher testing capability than others. For the ECDC it was hard to interpret this data and make decisions. But what decisions? Yes, the ECDC would be officially entrusted in drawing a collective European response to an infectious disease, but the problem is that this agency was never given any power. No other European agency has ever been crushed like the ECDC, from any side. The European Commission wanted the ECDC to be a merely consulting body, bringing on pure, unbiased scientific data from which the European Commissioners would draw their political decisions. The member states never wanted a European health authority stepping over their public health sovereignty. This sentiment is pure madness. The President showed that you can't mix nationalism with public health. A virus will never give a fuck about your sovereignty. Among all the anti-EU actions the national governments carried over the years, depriving the ECDC of any authority was the stupidest one, a true childish act of vanity and pride.[83]

The first agency director, Zsuzsanna Jakab, decided to set the agency HQ in Stockholm, in an attempt to be more independent from Brussels. Safe in the Swedish capital, Jakab used to directly contact national government officials, in an attempt to make them understand first hand the importance of a central European body which could help all the states to

[83]https://www.politico.eu/article/coronavirus-flaws-with-european-centre-for-disease-prevention-and-control/

coordinate together in case of an epidemic. Obviously, the European Commission didn't like this, as they saw Jakab's action as an attempt to step outside the strict boundaries the ECDC was given. In reality, Jakab just wanted the ECDC to become what it had to become: a beacon for all the European states, which would have projected its light to reveal all the possible infectious threats ready to hit the continent. A coach to lead the European players to win the game, the day a virus would have challenged the continent. Unfortunately, the EU is indeed democratic and putting public money on a body set to watch a threat that will "never" come is not so appealing to the average voter. How do you explain to people that you are going to pile up around €10 billion just to prevent influenza? It's understandable. I don't want to blame you here. Nobody actually thought a pandemic would have ever come. Even who expected it, didn't imagine that the economic damage would have been far higher than €10 billion. Now we know and I hope voters will understand the importance of increasing the ECDC budget. For the year 2021, the ECDC budget rose to about €63 million, but it's still insufficient to grant the safety of the European citizens.[84]

The political battle to raise the importance of ECDC is still ongoing. The risk this issue will be forgotten together with the pandemic is high. This battle is part of a bigger war that is being fought in the rooms of politics. Will the healthcare system come back to the center stage of public spending...?

Improving the Healthcare System

All over the world we need to improve the healthcare system dramatically, so that every country can be prepared to detect and contain a possible outbreak. Because Covid-19 is not the first outbreak and it won't be the last one. According to Alanna Shaikh, public health consultant, virus outbreaks are a

[84]https://www.ecdc.europa.eu/sites/default/files/documents/ECDC_Budget_2021.pdf

phenomenon destined to occur more and more in the near future.

She makes clear that no national healthcare system in the world is ready to cope with an outbreak of this scale, not even in the most developed countries. "Covid-19 showed all the holes in our healthcare systems...".

I invite you to check her TED Talk, which is an oasis of genuine information in the internet ocean of lies![85]

And please, follow her boring, but effective advices:

Wash your hands, wash your hands, wash your hands!

Sanitize your phone, which is one of the dirtiest things you carry with you!

Don't take your phone to the bathroom, man! It's simply disgusting...

- Don't touch your face,
- Don't rub your eyes,
- Don't bite your nails, for fuck sake!
- Don't blow your nose on the back of your hand... Jeez! Are there people really doing that? We deserve extinction then!

One last point she delivered during her TED Talk deserves a correction. She said:

- Don't wear face masks! Those are for sick people and health care providers. Since we saw during these days that these masks are limited, leave these precious tools to who really needs them.

This made sense at the beginning of the pandemic, when there were few face masks around the world. Once production scaled up, governments ordered their citizens to wear masks anyway, regardless of their condition. When you have enough units, the better-to-be-safe-than-sorry approach is the best.

The healthcare system must be improved in every country and this improvement, like any other, must start from you! At

[85]https://www.youtube.com/watch?v=Fqw-9yMV0sI

the next elections, ask your candidates what their plans for the healthcare system are. The local elections can have even a bigger impact. If every city and province puts more attention to public health prevention, central governments will have to follow suit. Remember how governments spend too much for the military but not enough for the healthcare....

"Ask not what your country can do for you - ask what you can do for your country!"

– J F Kennedy

Innovation Boost

Welcome to 2025! Yes, this is the year the world was taken to by President Corona. He boosted innovation so much that we moved forward by 5 years. There was only another time innovation was boosted so much: during WWII. President Corona boosted innovation with many less casualties, most of which already lived a fulfilling life, while he spared all the children, unlike the war did in the 1940s. Crises are the best innovation boost ever. When you face a crisis you have only two choices: you either innovate or you disappear! Fair and square! President Corona offered businesses these two choices. Luckily, most chose the former. Crises boost innovation because they present an array of new problems in need of a solution. Most problems were already present before the pandemic, but we needed the President to allow these problems to be listened to. Think about mRNA vaccines. This incredible technology has been in the pipeline for 15 years, but it was pushed and rolled out only when Covid-19 appeared. The solution was there, but the problem wasn't loud enough because infectious diseases affected mostly poor countries. Now that mRNA vaccines are a reality, we hope these will be used to fight and defeat gruesome plagues like Ebola or Dengue.

ZipLine saw its drones manufacturing business skyrocket.

With social distancing and lockdowns, their drones proved to be a safer, faster and more effective way to deliver vital goods like medicines, blood and vaccines. The drones are now being deployed in Africa, where even without Covid restrictions, delivery is a huge challenge, due to the long distances, dangerous areas, lack of roads and villages which sometimes are simply too remote to be reached fast enough. Drones can overcome all of these obstacles and deliver emergency material right in time.

If this is not awesome, I don't know what else is![86]

When it comes to innovation, you must mention the MIT of Boston. The geniuses came up with the idea of using UV light to kill SARS-CoV-2 particles. To test this theory, MIT created a cleaning robot and put it to work at the Greater Boston Food Bank. The robot covered a 4,000 square feet area within 30 minutes. It neutralized around 90% of virus particles and not only SARS-CoV-2. This technology is great because it provides a faster way to disinfect sensible areas, like a place where food is stored.

Current sanitizers are very effective at killing viruses and bacteria, but they can kill other life forms too. When sanitizers are washed away, their environmental impact is huge. Sanitizers are poisons after all. Unlike Trump says, it's not a good idea to ingest them, nor it's a good idea to throw them into the sewers after flooding the street with that shit, like they did in China. If you want to disinfect, from now on, use UV light, motherfucker![87]

If you can't visit a venue in person, maybe you can do it online. President Corona boosted the employment of tools like 3D reproductions, 360° views and VR settings. The result is important especially for museums, which can deliver their treasures with the best realism possible. Many could see

---

[86]https://flyzipline.com/

[87]https://www.covidinnovations.com/home/01072020/new-mit-robot-using-uv-light-to-kill-covid-could-be-used-to-disinfect-warehouses-schools-and-offices

museums they would have never visited otherwise and the magic is that, once saw in VR, they will be eager to visit the venue again, in person!

No other leader spread culture like President Corona. Leaders usually are not so keen on educating their own people. For them, people should remain ignorant. President Corona has a different view on the topic. That's why he brought us the amazing opportunities of e-learning. During the pandemic, we saw that home-schooling is possible and effective. Mind you, home-schooling should not become the norm. Kids need to socialize at school. Homeschooling, though, is demonstrating to be a powerful complementary tool for children when they cannot attend school. And even when they can, providing students with some flexibility, teaches them independence and self-drive, crucial qualities in the post-Covid job market. E-Learning also brought up a crucial issue. Not every parent can afford to buy a laptop or tablet for their kids. President Corona wants to urge governments to treat access to the internet and to electronic devices as a fundamental right for every student, regardless of the financial means of their families. This is a mission everyone should take care of!

Be the Change You Want!

I said already the world needs to be reshaped. This pandemic is the right occasion to rethink the whole world and the whole society. This crisis for sure has been depicted much more serious than it is in reality, but that's still good! It gives us the excuse to make those changes we want! Because the only way to change the world is to change ourselves! I hope you got the whole point through this whole book. You hate the world because it's unfair, violent, ugly, corrupt, selfish, greedy, complicated, hostile, twisted and maybe even dumb, but... these are traits which also dwell in you, in me, in us!

Sigmund Freud said that if you hate something or someone,

usually it's because it reminds you of something that you hate about yourself, subconsciously... So if you want to improve the world, work on yourself first! That's what I tried hard since the start of 2020. **Be the change you want to see in the world!**

2020 triggered in me such humanitarian spirit that I didn't have before. Sure, I've always made donations but they used to be small and happened seldom. That year I took from my wallet several times, to help against the COVID-19 emergency, but also to relieve the pain of Lebanon after the horrible explosion of the Beirut port or to give poor children a meal through the *ShareTheMeal* app. I even volunteered and that was the first time I did some real volunteering work! And I can see around, that I wasn't alone. Many people caught the message from President Corona and became more empathic, finding a higher purpose in their lives. But I'm aware that many of you reading this book are just lazy and selfish bastards so you might not understand and appreciate the beauty of 2020....

Luckily, my friends are not bastards like you. I was proud to see that many of my buddies saw the great occasion here and didn't think twice to join me in a special mission with the Food Bank. Since the day the situation looked critical, I wondered what I could do for my own community. Changing the world starts from your doorstep. It's here that my colleague Julien offered me to help the Malta Lifeline Food Bank to deliver supplies for its customers. The Lifeline Food Bank offers short term relief to unfortunate people, before structural aid comes from the government. As explained in their website https://www.foodbanklifeline.com/:

"Sometimes it can take a while before long-term support kicks in for those in need. Whether it's a result of a benefit delay, low income, ill health, homelessness, sickness or housing issues – we are here to help in the short-term.

When a Care Professional identifies a person or family who are experiencing a crisis, they are referred to us and we help by providing them with a non-perishable food-pack that can sustain them for one week."

In normal times folks go directly to the Food Bank premises to collect their supplies, but April 2020 wasn't normal time. We were asked to fill a van with boxes full of canned food, milk, cereals, water and drive this van around Malta to the addresses we had in the daily list. The mission looked simple, but even dangerous. PokerStars could not officially sponsor the initiative as we couldn't put our employees at risk of infection, but I knew that everybody at the office was ready to support us morally, should we have taken such a risk. I started the mission with my colleagues Julien and Natalja, but soon we recruited my friends too as the demand was increasing. I was driving 3 times a week, going even to places on the island I've never been before. It was an outstanding journey but also demanding. The most stressful part was the dirty tricks of Google Maps. Mountain View doesn't know Malta well yet and it tilts easily in these crazy roads and tiny one-way streets.

I remember with particular pleasure the company of my friends Andrea, Anna and Polina who served as excellent navigators and then Veronica who was also able to relieve my driving duties. 5 months after the completion of our adventure, Veronica moved my heart with this Facebook post:

Veronica Vai
November 16, 2020 ·

We are now 11 solid months into this world chaos. If you are not working/not getting a paycheck/struggling to make ends meet and run out of food or necessities...please don't let yourself or your kids go to sleep with an empty stomach. Please don't be afraid or embarrassed. Message me in private and I will drop groceries at your door and go. What's understood never has to be explained. Many of us have been in a tough position at some point in their life. It doesn't define you. But we have to support each other through this hard time. I don't have much, but I will support you with what I can. Veronika x

#JoinTheCause

You, Natalja Ja and 26 others

1 share

Friendship is among the most important aspects in my life.

Without friends life is not worth living. Thanks to President Corona, I could appreciate friendship more, in new ways that were so obvious yet so overlooked.

Before I used to meet a lot of friends in front of an alcoholic drink. Now I meet them in the most laid back fashion, usually at home while eating, petting the dog, watching a movie or simply talking about the most diverse topics! This helped me get to know them better, explore their least visible sides, go beyond their nightlife personas, appreciate their real selves!

We believe we know many people, but in reality, we just know many personas and only a few persons.

President Corona's Accomplishments

We can say it loud! This virus has superpowers! This virus is the first ever president of the world! We can see already that this president has accomplished feats impossible for any human leader!

There are many legends about President Corona. Whether they are real or not, it's up to you. For example, people believe that President Corona knows Victoria's secret! Other legends about the President say:

- President Corona can kill two stones with one bird.
- President Corona makes onions cry.
- When President Corona enters a room, he doesn't turn the lights on, he turns the dark off.
- When President Corona plays dodgeball, the balls dodge him.
- When President Corona watches the video of The Ring, the phone doesn't ring.

- Freddie Krueger is afraid that President Corona might one day get tired and take a nap.
- Some kids piss their name in the snow. President Corona can piss his name into concrete.
- President Corona counted to infinity. Twice.
- President Corona can divide by zero.
- President Corona can find the end of a circle.
- President Corona refers to himself in the fourth person.
- Bigfoot claims he saw President Corona!
- The saddest moment for a child is not when he learns Santa Claus isn't real. It's when he learns President Corona is.
- And remember, during summer, Jaws stays on the beach when President Corona swims...

It's just thanks to President Corona that summer 2020 was an outstanding season. Crystal waters, clear beaches, quiet cities, without noisy and harassing tourists, no queues, no traffic-jams and no smog from unbearable honking cars, and above all, clean air! Whether you were wearing a mask or not, you could feel the difference!

Cleanliness is the first result that was visible and tangible about the President's disruptive mandate. Astonishing results which could be observed even from space!

Cleaning the Air We Breathe!

We all have seen the spectacular satellite photos of China that showed how the country got much cleaner after the general halt of industrial activities.

President Corona could be the very messiah that the greens were waiting for, even more effective and categorical than

Greta Thunberg herself!

In the first 3 months of 2020, President Corona did what no other green policy has ever done in decades! President Corona started solving the never-ending climate crisis everyone complains about. This little buddy is the very cure for the planet! Just stop all the industry activities et voila'! Problem solved!

Well, even here your superficial mind is tricked once again. The air pollution in the sole region of Beijing dropped, but not as sharply as you might think. It dropped by 25% compared to the same period of 2019. Remarkable, but not enough to solve the climate crisis once for all.

All the newspapers mention this report by the independent agency CREA as the demonstration that Covid-19 was cleaning China from its oppressing pollution. It was, but only in part. To understand why, it's important to read some passages from this report:

- *The two largest sources of air pollution in the Beijing region are steel industry and the heating of buildings. This contrasts with the regions that have seen the largest air quality improvements, where power plants, transport and smaller industries are more prominent. These two sectors are the least affected by the virus – buildings need to be heated even when cities and villages are locked down, and might even be heated more as people spend all of their time indoors. The steel industry is loath to shut furnaces as shutdown and startup incur a lot of costs. So the industry has kept on producing, resulting in record-high stockpiles.*

- *Beijing's geography makes its pollution levels very sensitive to wind directions – enormous industrial clusters and high population densities to the south and east mean high pollution when wind comes from those directions, while wind from grasslands to the north and west brings clean air. So bad luck in terms of wind direction and speed can have a major impact on week-to-week pollution levels.*

Another great opportunity, but as you can see reality is never that simple. Nevertheless, President Corona's work in cleaning the world has been impressive. Let's take a look at other major regions, where traffic is the most prominent polluting factor.[88]

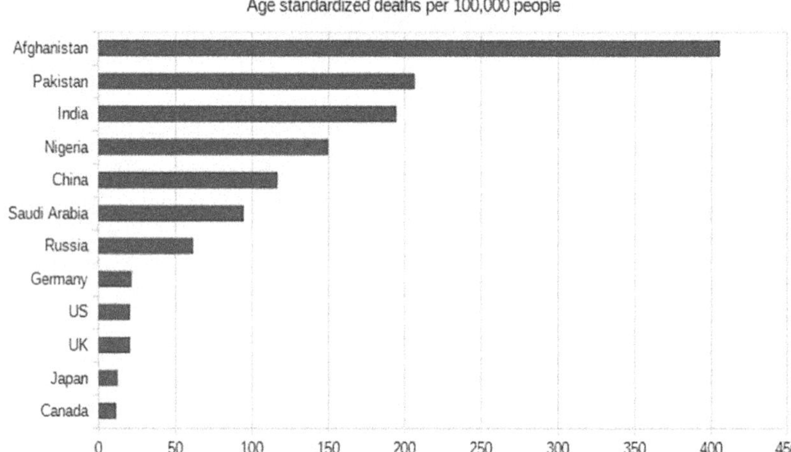

The stats above are from 2016, but there is no reason to believe in 2020 the situation was any better. Even though air pollution seems to kill less than COVID-19 on average, we need to consider that air pollution is constant. It's been there for a century and it seems bound to stay with us for long still. No one is fighting air pollution with the same effort we were fighting COVID-19. The only one who did something concrete about air pollution is President Corona himself, who you are trying to kill so desperately.[89]

Of course, the swift cleaning brought up by the lockdowns

---

[88] https://www.statista.com/chart/13575/deaths-from-air-pollution-worldwide/

[89] https://www.weforum.org/agenda/2020/04/the-deadly-link-between-covid-19-and-air-pollution/

can do nothing against the issue in the long term. The President just started a work which must be taken over by human leaders. We all could enjoy first hand how clean air makes us feel better. Focus on those days from March 2020. If you couldn't go out, you could still peer out on your balcony. Remember those breaths... Yes, focus on those. Wouldn't you like to breathe that same air again, forever?

Nature Took Space Back

We will never forget when dolphins visited Venice. That was an incredible show. A show which was staged even in many other parts of the world. City centers got void of humans and after a few days of lockdown, animals took the courage to visit those areas. Deer, boars, goats, but even buffalos, wolves and bears came into town. It was epic! For nature, the virus was an effective tool to push humans into their homes and allow animals to regain territory, after centuries of unstoppable receding. The human kind has always been the bulliest species ever appeared on Earth. Humans destroy and kill every form of life they encounter while expanding their never-ending vital space. This Nazi pattern doesn't seem to slow. Ask your grandparents how your town used to be just 50 years ago. If you live in a suburb, they will probably tell you how nothing was there. Your house didn't exist. It was just fields and woods. I always think about the symbol of Madrid. It's a bear eating fruits from a tree. The reason is that the countryside around Madrid was inhabited by many bears. In the 21st century though, you can't find a single bear in that area. Just humans, humans and other humans. Some cats and dogs, sure, but they count like humans when we relate to wild animals. Like humans, even cats and dogs have huge demands and consume an unsustainable amount of resources.

All the wild animals who get in contact with humans, get either killed or imprisoned. The President was ruthless in teaching us how bad isolation can be. Being confined in a few square meters can drive you crazy. It hurts even more when you undergo such conditions after you know you did nothing wrong. I hope the President's lesson can make humans open their eyes on the many animals they caged for their own entertainment. Leaving alone animal experimentation labs, think about all the zoos around the world that are still open. The only things open in these absurd places should be the cages. Maybe the President gave you a taste of what those poor creatures experience every day just for the enjoyment of your sadistic family.

Because President Corona is not here to kill the weak. He is here to stand up for them!

Tackling the Housing Crisis

You were stripped by the joy of traveling. That's one of the most "brutal" things that President Corona did to you. But he did that with a higher purpose. Because you no longer showed up in hotels, those premises could be turned into shelters for homeless people.

The President offered thousands of rough sleepers around the world the chance of having a safe place during winter, especially in places like New York and London, where this season is unforgiving. When you have a comfy place, that satisfies all your basic needs, you can have a better rest, hence you have more energy to find a job, perform better at that job, show up clean at job interviews or even study with more piece of mind, when you have internet access and a quiet place for your own. In other words, President Corona gave these homeless individuals a precious chance to get integrated into society and get what all the others have.[90] [91]

---

[90] https://www.economist.com/1843/2020/08/07/homeless-at-the-holiday-inn

[91] https://news.stlpublicradio.org/government-politics-issues/2021-03-25/st-louis-moved-homeless-people-into-hotels-putting-some-in-danger-we-were-an-eyesore-to-them

In Barcelona, where all the houses were turned into Airbnb, locals could finally find a room in the center, again! Finally the locals could live in their city again! The local government liked the situation so much that it banned Airbnb forever from the city center! The crackdown comes for rooms. They won't be able anymore to sublet private bedrooms on Airbnb. The only kind of listing allowed will be entire apartments and only with a regular tourist license issued by the local council. Of course, these licenses are near-to-impossible to get. President Corona stole from the rich to give to the poor![92]

Similar initiatives took place in Lisbon, where the housing crisis is soaring. Lisbon is the new go-to place for tourists and digital nomads. No wonder the local landlords turned to Airbnb, pricing out the locals who used to live in Alfama. Now, President Corona reminded them that leaving some property available for a long let could be a good idea. You ease the risk of dry tourist seasons. The local council even started a campaign to rent empty houses for a lower price but with the guarantee of a 5-year rental agreement. The local council will then sublet those properties at a fair price, without speculating.[93]

Airbnb is one of the most innovative businesses arisen from the 2008 financial crisis. It was a massive opportunity back then, when most people had lost their jobs and needed to create alternative sources of income. On the other hand, tourists felt the need to find cheaper solutions than classic hotels. Here is how Airbnb combined the two needs together. The problem came when Airbnb began to demand higher and higher standards from its hosts. Soon, private houses became just like hotels, including the price. Hosts turned into proper hospitality entrepreneurs, so they began to buy and turn every house in the city center into a boutique hotel. Landlords

---

[92]https://www.forbesindia.com/article/global-news/barcelona-takes-on-airbnb/70571/1

[93]https://www.theguardian.com/world/2020/dec/01/covid-created-an-opportunity-lisbon-turns-20000-tourist-flats-into-homes

followed suit. Renting for short term on Airbnb was too much more convenient than letting for long term to students and workers. What losers! Why don't they come here just for a few days to enjoy life rather than studying and working! Of course, Airbnb is not at fault. Airbnb created a win-win situation, then institutions failed to regulate the new market in time. Politics systematically ignores the best medicine practice: prevention is better than curing. It took President Corona to fix the issue and even him could do just a partial work. There is still a lot to do in order to solve the housing crisis. Too many cities have houses which rent is as high as the average salary.

Is it possible that today an experienced worker in his 30s has to share a house with strangers? It looks like the *Kommunalka* from the Soviet Union. Kommunalka were communal apartments shared by several families together. Families used to have one room each and shared the kitchen and other services with the other families. It's ironic how today they label you as a communist the moment you protest against prohibitive rentals and demand regulations for the rental market. They want cities where people are confined in their rooms without any other private space, owining nothing, just like in the Soviet Union... Who's the real communist?

In Canada, the province of British Columbia, seized the moment to purchase two hotels in Vancouver for a total of 173 rooms, to house the most vulnerable of the community. It's a generous act of social housing, in a city which became prohibitive for most.[94] Similar initiatives are setting foot in every part of the world. President Corona wants to make clear that a house is a fundamental human right!

---

[94]https://archive.news.gov.bc.ca/releases/news_releases_2017-2021/2020MAH0070-001160.htm

Remarking the Importance of Our Relationships

I clearly remember how many complained about the little time they spent with their families prior to Covid. Most parents work at least 8 hours a day, plus 2 hours of commuting. That means they are away from home for at least 10 hours a day. If you add up all the tasks that an adult might have to complete, take out another 2 hours a day. From this, you need to take out an additional 8 hours of sleep. This means that on average, people see their kids and their partners only 4 hours a day. That's insane and makes me wonder, why do people make kids if they are never with them?

President Corona forced parents to spend full time with their kids. The reactions were mixed, but I like to think that most parents, in their depth, thanked the President for this precious opportunity. I live in a big house that I rent from my buddies James and Karolina. They live upstairs while I hang at the first floor. Yes, we live in a Kommunalka, OMG!

We spent a lot of time together during the early days of the pandemic, when we were afraid of going out even for groceries. Their kid Leo was there too, spreading nothing but his child energy. Like every kid, Leo wants to play all the time, jump, shout, laugh. I understand that for parents this can be exhausting. I remember one evening Karolina complained, wondering when childcares would have reopened. I comforted her by saying: *"I know you are looking forward to it, but one day you'll look back at the good old days, when you had the privilege of spending so much time with your kid...."* she smiled and nodded.

I still believe this. I even believe that in the years to come, there will be a sense of pandemic-nostalgia. There are folks feeling nostalgia for ruthless dictatorships which brough no good at all. Why shouldn't someone feel nostalgia for two years full of unique opportunities?

Taking Care of Our Health More

A pandemic is a great occasion to remind the importance of taking care of our own health. Indeed, the Spanish Flu pandemic marked the rise of an important fitness discipline still popular nowadays: Pilates! This discipline, so similar to yoga, derives from Joseph Pilates. Mr Pilates was a complete athlete. He was born in Germany, as a fragile child affected by asthma, rickets, and rheumatic fever. Rather than getting depressed, he dedicated his life to improving his physical strength, a practice he turned into a life-style. This drive led him to explore gymnastics, boxing and martial arts. Beside this, he developed new ways of training his body, with minimal equipment, that soon attracted the attention of many students. When he moved to England, War World I broke out, and as a German citizen in the UK, he was interned. During the prison time, he honed his training techniques, helping his prison mates to get stronger and healthier. It was during this time that the 1918 Influenza Pandemic broke out. According to contemporary recordings, none of the Pilates' students died of flu. That was the rise of the discipline he called "Contrology", which a tiny part of is today marketed as "Pilates". Contrology is a wider spectrum, it concerns the whole life-style of a person. Back already in the 1910s, Joseph Pilates concluded that most of the diseases affecting the Western society were the result of the weary modern life-style that brought damaging habits, such as drinking, smoking, bad posture, junk food, lack of healthy physical activity and pollution. Does it sound familiar? It sounds like 100 years later we are still stuck with the same problems. Elderly alone, who among the young population is hit most by COVID-19? Who has bad habits of course. But just like the Spanish Flu made people discover the modern concept of fitness thanks to Pilates' new approach, even COVID-19 is shaking people's awareness about their own health and the role their will plays on it. Most people can decide to be healthy.

The finding that smokers have 40-50% higher chances of severe COVID-19, motivated millions of people around the world to quit smoking. In the UK, just in the first half of 2020, 1 million people quitted smoking. That's impressive. I know that some other studies claim smoking increased during the pandemic, but those were occasional smokers affected by the anxiety a pandemic can provoke. In the long term, people understood how important it is to take care of their lungs, regardless of Covid. The lungs allow us to do the most basic act in life: breathing! How can we even consider of damaging them?

However, as even reminded by the WHO, smokers need assistance to quit tobacco properly and avoid falling for it again.

"Currently, over 70% of the 1.3 billion tobacco users worldwide lack access to the tools they need, and the gap in access to cessation services was further exacerbated in the last year as the health workforce was mobilized to handle the pandemic."[95]

Beside calling everyone in a collective effort to fight tobacco, the WHO created an amusing AI, named Florence, which works as a virtual assistant to help you give up your cigarettes. She still sounds quite robotic, but she's cute. Try her for fun at least![96]

If the pandemic had a positive impact on tobacco consumption, we cannot say the same about the worst drug ever: alcohol. Drinking has been on a steady rise since the start of the pandemic. Even who wasn't accustomed to drinking much, started to open a beer even when alone at home. Maybe some used a Zoom meeting as an excuse to drink at home. I confess it happened to me as well. I've never ever drunk alone and then I found myself cheering over Zoom.

According to a global study, alcohol consumption in 2020

[95] https://news.un.org/en/story/2021/05/1093102

[96] https://youtu.be/R8Uw0D9Ytso

increased by 14% in the US, 21% in Canada, 17% in the UK, and 25% in Australia. For Europe the data is fragmented among states. Belgium increased by 21%, Poland by 14.6%, while Greece kept on the same levels of 2019. Good job! Probably Greeks are more coached against crises![97] [98] [99]

Here President Corona made fun of us humans, so desperate for a hydrocarbon which gives us the illusion life is better.

We rely on alcohol to have "fun", though nobody can explain where the fun is in drinking. We need alcohol to get laid, just to sleep with who we don't even like. We drink because we think alcohol makes us stronger, maybe even invincible! The truth is that the very act of drinking is a strong sign of weakness. I appreciate drinking some quality wines, quality beers and quality Japanese whisky, but I hate myself when I indulge in cheap drinks just to stay tipsy. I understand the temptation. When you're tipsy, you feel like your brain is flying, hovering over other people's thoughts. You start speaking more fluently, without any hesitation. You might even become capable of speaking a foreign language you just studied on DuoLingo! This is great, but then as you feel this nice sensation fading away, you feel the urge to "refill". Your memory skips directly to the next morning and that strong headache we call hangover. Drinking is the art of borrowing happiness from tomorrow...

Quitting smoking and drinking is not enough, if you don't start exercising too. You don't necessarily have to lift heavy weights. Even a daily run is a great boost for your health. Running is the opportunity many seized during the lockdown. As many jurisdictions excused runners from home arrest, a lot of people started to run. Some citizens began to target runners as if they were black plague spreaders. Some even called them "murderers" as if they could spread the virus in the air. I remember an Italian mayor who burst out of anger yelling:

"Now everyone is a runner! The fuck are you running for???

---

[97] https://www.ncbi.nlm.nih.gov/pmc/articles/PMC7944101/
[98] https://pubmed.ncbi.nlm.nih.gov/32503173/
[99] https://www.ncbi.nlm.nih.gov/pmc/articles/PMC7763183/

Until the other day it was just me and two other dudes. Now the entire town is running marathons!!!"

At least, people understood the importance of sport and the freedom it can give you. Please, keep on running even after the pandemic ends. Being fit reduces the chances, not only of COVID-19, but of any respiratory and cardiac disease. Sporting means less pneumonias and less strokes, put in layman's terms. Sure, there were many fitness freaks who died of COVID-19, but many more were those who survived or were even asymptomatic.

The last piece for your best health is to follow a balanced, healthy diet. I know it sounds banal, but that is. We are what we eat. If you eat shit, you'll be shit. Very simple. Drop your disgusting hot dog, throw away any ketchup and mayonnaise from your dirty fridge, if you don't want to become an easy prey for President Corona and other deadly pathogens. Mark Rothkranz says that viruses and bacteria are not the bad guys, they just kill who is already sick. They kill those who didn't love their own bodies and didn’t take proper care of them.[100]

About diet, I see another golden opportunity. President Corona taught us how important Vitamin D is and how little of it we get. We are closed in our offices for too long. We go out more and more only during the night as if humans were owls. This is an invisible lockdown we have been inflicting on ourselves for a century!

This should make us consider reshaping our ways of working. Not only having offices with more sunlight, not only having offices outdoors, but we should consider creating or resume outdoor jobs. For example, we should promote the farmer role again. That's actually a cool job that requires a lot of dedication, organization and planning. A good farmer has nothing to envy from a Project Manager, with the difference that farmers' work serve society more than many empty businesses which just create artificial needs.

---

[100]From Markus Rothkranz – Heal Yourself 101: Get Younger & Never Get Sick Again

Farming should be incentivized again in the West. It would take people out from office, where many don't fit and never find their purpose. More people would have higher levels of Vitamin D and future respiratory pandemics will have a lower impact on hospitals. With many little independent farmers, we could prevent billionaires like Bill Gates from controlling most of the food production. Many little independent actors who produce local food. It's something that even your friend Russell Brand talks about often. There are many opportunities today for farming. You can grow all the classic products, but locally. People love local products more and more, not only because they taste better, but also because their transportation has a low impact on the environment. Carrying strawberries from one village to a near town is less polluting than importing them from the other side of the world.

Another great opportunity is given by the legalization of cannabis. I see already around me a lot of people taking care of their marijuana plants, even when they had zero interest in gardening before! Even here, a lot of people trapped in office life may find freedom and a sunny intake of Vitamin D with Mary Jane!

Jobs Created

If farming could be the future, there are jobs which President Corona created right now, during the chaotic days of the pandemic. If you are still unemployed, you may want to check the following sectors we are going to discuss.

Healthcare sector

What was the whole point of lockdowns around the world? The point was our healthcare system cannot cope with the number of patients the coronavirus is capable of sending to a hospital. We didn't want to see what we witnessed in India during the rise of the Delta variant, when people were stocked in the streets, waiting for their share of oxygen. Even without this dramatic example, the more modern hospitals in the West were put under pressure, with doctors and nurses forced to impossible shifts in the ICU, covering even 16 hours straight. You understand that working 16 hours in a row is not sustainable and indeed, too many healthcare professionals started to quit their positions as of late 2021. Enough is enough!

The good news in this is that institutions started to realize how healthcare professionals are not important for their society. They are vital! Without them, death would be part of our everyday life, with bodies left in the street, just like it used to happen in the middle ages or just like it happens in some parts of India even today. I have nothing against it. I'm not afraid of death, but I wonder if you would be OK. Our society made death enemy number one, something to avoid at all costs. We wanted to remove death from everywhere, even from our language. Think about it. When a celebrity dies, the media rarely say "He died". They will use expressions like "she passed away, he left us, he's in a better place now" and other clean terms of the same kind. You don't like death? Then ask your government to increase the salaries of your local healthcare professionals and to improve their working conditions. When you work 16 hours a day, it is no longer a job, it is slavery.

Now, with many healthcare professionals fed up with their jobs, there are many vacancies left. You may want to consider studying medicine. Hospitals also have many positions for para-medic staff, which don't require a whole, lengthy university course. Since you feel so tough that a virus will never kill you, you might put your superpowers at the service of the healthcare system. Think about it, Superman!

Mask producers

Everyone can be a mask producer. The surge in face masks demand scaled up a business which used to be a niche. Basically, only surgeons, dentists and their patients needed a face mask in the pre-Covid world. Now, everyone needs a mask! They need hard working people to produce them! They are hiring! Maybe it's not your dream job, but when you lose your employment overnight, sewing face masks is not that bad. At least some manual work which breaks the office routine you were too much used to. There is nothing shameful in working with your hands. In the West, manual labor has been seen with more disdain each day. This mentality had tremendous effects, especially in Europe. More than America, Europe dropped most of its manufacturing industry to move it to Asia. Lots of jobs burnt, in return for cheap, but low-quality items invading our shelves. Many of these items turned out even toxic and dangerous, for us and our kids. President Corona forced Europeans to rediscover manufacturing and the nobility behind it.

I remember once, when I was 18, I was walking with my buddies and we bumped into my former school mate Fran. He proudly said he had found a job as a welder. I was happy for Fran. I greeted him and then I kept walking with my buddies. After a few meters, the guys said, with disdain in their voice: "Meh... welder..."

I was irritated. What's wrong with being a welder? In their opinion, everyone should aim at doing non-manual jobs. It seems like an office job is the only noble aspiration one should have in our tertiary sector society. Yeah, a job which will make your legs tiny and your belly huge... Great aspiration! I've always had a love-hate relationship with office jobs. I love the creativity part of an office job. You're in an office to create, to share ideas and execute them. On the other hand, an office job can be detrimental for your health, much more than being a welder. I always suffered sitting for long hours. Once, I even had to undergo a surgery, as my ass developed a thrombosis, which resulted in the most painful experience: hemorrhoids. It took me months to get in shape again. After that traumatic experience, I invested in an ergonomic chair, but I also started using standing desks. When you spend more time standing, you immediately feel the benefits. I started to have more energy, I needed less sleep and my mood was definitely better. My advice is don't discard manual jobs if you bump into one. If you don't want one which you think may stain your CV, consider doing some voluntary work. It's an eye opening experience and free gym! Even the most humble voluntary work makes your resume 50 times brighter! Recruiters will love you!

If you had an experience as a mask producer, don't be afraid of putting it in your CV. In that moment, you were a hero, helping your community with mass protection. You must be proud! You can even put "superhero" in the job title. I'm not kidding. You were! Plus, this weird title will catch the recruiters' attention and will be a good starting point for a bubbly job interview, which will make your application memorable!

Sanitizer producers

Another industry capable of absorbing the layoffs sparked by the lockdowns was the chemistry industry. Hand sanitizers needed workers more than ever. The global hand sanitizers market was valued at $1.64 billion in 2019. In 2021, the market reached $11.4 billion! A +595% leap forward![101]

Though most of the hand sanitizers production takes place in India, distributors employ people around the world. We can bet the hand sanitizers distribution employed even those people left without a job in your neighborhood. For every catastrophe, there is an opportunity.

---

[101]https://www.arizton.com/market-reports/hand-sanitizer-market

Some people went even further to become private hand sanitizers producers, crafting special, handmade products. A great example is Handcrafted In Laurelcanyon, an online shop which sells a natural, handcrafted hand sanitizer. Hands down! OK, enough![102]

This is a brilliant idea, especially for those who don't trust what chemical companies put in their products. Sanitizers can even be highly polluting, so opting for a natural one can be a winning solution to preserve the environment while you keep safe.

E-commerce

During a pandemic, beside masks and sanitizers, there are many other items you can sell online. You can sell everything online. Maybe the loss of your job was that wake-up call you were waiting for. In your depths, you knew you were bound to start your own business. The best and easiest business you can start nowadays is an online business. E-commerce was the fastest growing sector in the pandemic. Shopify doubled their revenue in Q1, with a striking +118%. Shopify is a Canadian tech-company which democratizes e-commerce. They offer you turn-key solutions for your e-commerce website, without having to hire an expensive developer. They have countless modules which are proven to be effective at guiding your customers to purchase your products. Your online store on Shopify is highly customizable, fitting any product you think to sell. If already these features make Shopify superior to Amazon, the real perk is that with Shopify, you retain your customers database! You know who your customers are! This doesn't happen with Amazon. Jeff keeps all that essential data for himself and this puts Amazon in the position of stealing your business if they want to.[103]

During 2020, global e-commerce retail sales grew by 25% in

[102]https://www.handcraftedinlaurelcanyon.com/

[103]https://www.forbes.com/sites/laurendebter/2020/05/07/shopify-small-businesses-online-coronavirus/?sh=36372a73a64a

volume, topping $4.2 trillion worldwide. The impact of Covid-19 led people to use e-commerce even for purchasing everyday products like groceries.[104]

This means there was never a better time to start your online store! If you have an idea, go for it! The investment needed for an online shop is very limited, while the returns are limitless. So many people and families started a new life with e-commerce. Thanks to President Corona, they are now running that business they thought they couldn't even start. When you lose your job overnight, you either start your online business or you starve. President Corona's methods are brutal, but very effective!

Training and courses

For you it's hard to believe, but some of the people stuck at home during the first wave wanted to grow! That's how the demand for training courses exploded. Obviously, the biggest chunk was coming from who had lost their jobs and wanted to requalify. Microsoft even started a whole program in collaboration with LinkedIn to help unemployed people make their way into the tech industry. Others looked at Udemy, a place which created opportunities for both sides, the students and the teachers. Students could train on skills which are in high demand now, such as data analysis, blockchain, 3D art, machine learning. Teachers could finally monetize their skills. If you are confident you can teach something, it would be a good idea to prepare a professional course about your subject and post it on Udemy. This platform will help you create a passive income out of your own skills. It's simply awesome, especially when you are a professional who has lost the job due to the pandemic.

Online education can be an opportunity even and mostly for those who can teach a language. If you have passion to teach your language, you may consider getting a qualification and

---

[104]https://www.statista.com/topics/871/online-shopping/#dossier-chapter1

start offering your services on platforms like Verbaplanet or Justlearn. It's fun, easy and you can do this job from anywhere in the world! Don't think your native language is not in demand. People might want to learn a language for the most diverse reasons, not only for a professional purpose. Sometimes the most exotic your language is, the better! It's the easiest way to stand out!

The surge of online gambling

E-commerce is not the only online sector which boosted revenue and created more jobs in 2020. My sector enjoyed the reforms of President Corona more than anyone else. The online gambling industry doubled if not tripled its revenue, despite the closure of any sport event. People couldn't bet on sport events, but they could still bet on virtual games and e-sports, while horses kept racing. Then, casino games never sleep. That's where bets went on, while governments locked their people at home. Slots are stupid games which make you stupider. I know, but who plays those would have found something likewise stupid anyway. I prefer those who play poker. 2020 saw a resurgence of poker. At PokerStars we could feel like a poker company again. We were going back to the origins. It was a proud feeling!

Our workload skyrocketed, but we didn't mind when we saw our revenue for Q1 of 2020 going up by 92% on a YTY basis. It was a record! I was happy, because poker is a game everyone should play or at least learn. Should more people play poker, they would better understand why vaccines work on some people and not on others and would be able to gauge the risks associated with a pandemic, making better decisions. Online poker saw its heyday after 2006, when Casino Royale screened in theaters. James Bond is the best poker teacher ever. He's so charismatic that after his performance in Montenegro against Le Chiffre, everyone wanted to emulate him. Casino Royale saw the definite launch of Texas Hold'em into mainstream entertainment, 3 years after the victory of Chris

Moneymaker at the World Series of Poker (WSOP) Main Event, in Las Vegas.

Chris Moneymaker got a free ticket for the main event by winning a satellite tournament right on PokerStars. While the buy-in for the Main Event is $10,000, the satellite tournament played by Moneymaker was just $86. That's what he paid to play in Las Vegas. He ended up winning the Main Event tournament, for a total prize of $2,500,000, plus the WSOP bracelet which is priceless. The incredible journey of this former accountant and amateur poker player become professional, sparked the so-called *Moneymaker effect*, which is the hope that every amateur player might one day qualify to the biggest live poker tournaments in the world by investing an affordable sum online. Of course, no one managed to emulate Moneymaker, not in one shot at least. Nevertheless, online poker helped many talented people to make a living out of this noble game. Some even reached Las Vegas, though after years of working up the ladder. Then, as the Moneymaker effect and the Casino Royale fever faded, players lost interest in poker altogether. As it is a skill game, the players winning were more and more the same, until very few used to win most of the liquidity present in the online poker rooms. Amateurs got discouraged and they turned back to sports betting and casino games. The decline seemed irreversible, until President Corona came...

The surge of gaming

In April 2020, Twitch saw a 60% increase in its viewership, compared to April 2019. During the lockdown, people looked for new ways of keeping engaged and gaming was one of the most popular destinations. There is nothing better than a Call of Duty battle to fight boredom and forget about the virus! Videogames give you the chance to immerse yourself in another world and forget this valley of tears. In gaming worlds you have to abide by rules you chose and agreed upon. If you don't like them, you can change the game, hence the world.

About the real world, first of all you couldn't choose it. Your parents decided for you that you had to come into this dire world. Second, if you don't like this world, you can't change it. You still must abide by its rules, whether you like them or not. This will be valid forever. Don't think that direct democracy or decentralization will make it any better.

The surge of gaming created an array of collateral jobs. New gamers discovered they could make a living out of streaming their gameplays on Twitch! Out of the blue, those who were laid off could get a new income from their biggest passion! This is simply beautiful. Who was an already established Twitcher, could scale up their business. You see? Mr Corona wasn't bad for everyone. Mr Corona was even great for those who can catch opportunities. If you can't catch opportunities and see the dark in everything, you will never be happy, even if you found yourself amidst the brightest of the golden ages!

OnlyFans

For many loners, the greatest accomplishment of President Corona was to set the perfect conditions for the explosion of OnlyFans. Don't pretend you don't know what it is...

OnlyFans is that platform where creators can monetize their content by receiving monthly payments or tips directly from their followers. The content available ranges from fitness classes to exclusive cooking classes or even private concerts from notable musicians, but guess what the kind of content fans are most eager to pay for? Of course sex!

Founded in 2016 by a mysterious London based company named Fenix International Limited, OnlyFans was immediately a money machine, but it was only in 2020, when the President came, that the site broke into the mainstream culture. In 2020, OnlyFans saw its content creators increase by 40%, while its subscribers increased by 1,100%!!! Today, OnlyFans has something like 100 million active users!

What do these users pay for in practice? In our society

obsessed with celebrities and sex, OnlyFans combines these two obsessions in one place. Here, you can have the chance to see your favorite singer, actor or influencer engaged in sexual acts and have the thrill to see how they look like under their clothes. Or maybe, here you can find that sexy neighbor you've always been fancying about. But the gratification for your limbic system doesn't end here. When you become a regular donor, your favorite star might create personalized stuff for you, establishing a closer, interactive relationship. This is probably the real reason why OF got momentum during the pandemic. Many fans declared that OF provided an almost girlfriend-like experience in a moment when they were confined at home alone, without the chance of meeting any person in the flesh. It wasn't just the spicy content, but also the most innocent moments, like when your celebrity lights up the candles on a birthday cake singing your name. We always say that to succeed, a business must solve a real problem society has. During the pandemic, society had a desperate need for emotional connection and OnlyFans filled that gap extremely well.

OF also provided new income opportunities for those thousands of women (but even men) who lost their jobs while President Corona was reshaping the world. Sex workers of course had to reinvent themselves too as direct physical contact with clients had become too risky, while normal folks found an easy way to collect cash for their bills, selling their bodies in a clean way. This is what we can call a win-win-win situation. Fans win, creators win and OnlyFans wins even more. These three parties, more than anyone else, must shout: "Thank You, President Corona!"

Influencers

You may associate the word influencer with modern day social media. However, influencers are as old as humanity itself. The village shaman was an influencer, the king was an influencer, painters and sculptors were influencers, directors and singers have always been influencers. Social media just expanded the possibilities to be influential on something. Kyle Jenner is influential when it comes to make up and fashion, just like Ben Armstrong (aka BitBoy) is influential when it comes to crypto investments. They might not be relevant to you, but they are relevant to others. They have their niche and perhaps, you might find your niche too. You have influenced someone during your life too. Influencing and being influenced are the most natural acts we do, every day. Why not create an activity out of this? Like any job, influencers can be good influences or bad influences. It all depends on their content, their audience and the expectations of both sides. A great example are yoga gurus. They promote a healthy life-style, they share yoga workouts for free on YouTube, they post inspirational quotes. On the other hand, some of them (not all), share radical views, like mistrust in vaccines and modern medicine, while promoting "natural" supplements for which they get generous affiliate commissions. Their followers are angry at the profits made by Big Pharma, but they are OK with the profits made by their yoga gurus, by selling "natural" supplements which most of the time are not natural at all.

Nowadays anyone can be an influencer. Gen Z is exploiting this trend at its best. Most of the greatest TikTokers are from the Gen Z and thanks to their creativity, they've become wealthier than their parents, while working less or at least, working with a purpose. Of course, their parents are not happy with this. I could see, especially on LinkedIn, the disdain of many Gen X and Baby Boomers for how their kids follow "easy money". After destroying any social and work security for their kids, beside destroying their very environment, the old generations are not content. They even demand their kids to work for the minimum wage, without any right and any security, while paying a rent which will benefit an old big landlord, while they will never have their own house where to start a family. On LinkedIn, some pointed out that, if making money with TikTok gets too easy, kids will drop from studying important subjects like medicine, leaving the society of tomorrow without doctors. Well, first of all, not everyone is creative enough to make a living out of TikTok, but even if that was the case, does our society deserve doctors? President Corona unmasked our appalling lack of respect for the healthcare professionals. Why should kids pursue a career where they will:

- risk their lives,
- be underpaid,
- live under constant stress,
- face lawsuits from their patients' families,
- be called "paid actors" at the next pandemic,
- be shot by some lunatic who thinks they're killing children,
- hear their dying patients in the ICU "I'm dying because you forced me to wear a face mask!"

Would you pursue such a career? Would you save yourself if you could see yourself from outside? I don't think so. If you could see yourself from outside, you would unplug the ICU machines and let the computer utter that flat beep sound, the sound of the flat line...

To Gen Z I say: don't listen to your parents and grandparents. They are just envious. Don't rebel against your governments. Rebel against your parents. Your governments were voted and shaped by your parents! Take back the future your parents destroyed! And should the next virus take your parents and no one will be capable of running an ICU, that's OK! May they rest in peace...

Sparking Our Curiosity

YouTube videos like "what is a virus?" or "how coronavirus kills" and similar topics got the same views of a Billie Eilish hit. We can definitely say that the pandemic sparked the general audience's curiosity, pushing millions to study bits of medicine. Now everyone knows what a virus is, how it works and why we get it. The general public is now a little less ignorant and for this we must say Thank You, President Corona!

If you are interested in getting deeper about COVID-19 and virology as a whole, I suggest you these amazing YouTube channels who helped me a lot with understanding the dynamics happening with this disease:

Doctor Mike Hansen

Dr Mike Hansen has been my official COVID-19 guide. Hansen is an American doctor operating in the ICU during the pandemic. He's a true frontliner, so he knows what he's talking about. His battlefield stories are always backed up with data and detailed explanations, laid out in simple terms so you can understand them as well. To understand Covid, Dr Mike Hansen's channel is a must-watch! And when educated about Covid, this channel will provide you with insights on many other medical topics. I've learned how sleep works, how to have more energy, and what prevention means.

Medlife Crisis

Dr Rohin Francis is a funny doctor from the UK, whose mission is to bust the bad medicine advice, pseudoscience and quackery you find on the internet. He will take you into a journey where your critical thinking will be really put to the test. Leave your expectations at the door and be open to everything, especially sharp humor! A good starting point is Medlife Crisis video *How Athletes Fall for Pseudoscience, and Wellness Leads to Conspiracy Theories*. Yes, it's about Novak Djokovic and the Australian Open... I didn't write anything about them, because this whole situation is a shitshow, where both parties are at fault. Hence, they don't deserve a single line in this book.

Kurzgesagt – In a Nutshell

This is an educational channel which relies on super cute animations to explain complex topics with disarming simplicity. Ideal for kids, but even for their parents! Often, these topics rise from weird questions, like: what if the Moon falls onto Earth? Or, how unaware things became aware? How to terraform Venus? And many other mind-bending questions!

Their animated characters are even protagonists of compelling stories. Yes, this channel created even short films, with deep philosophical impact. My favorite is *The Egg*. It makes you think of death under a completely new perspective. I'm not going to spoil more... I just tell you that I cried with joy at it!

TEDx Talks

You're not a nerd if you don't have TEDx Talks in your favorite YouTube channels! What I love about TED is their content, which is so valuable, interesting and mind-bending. Every TED talk makes you think and challenges your knowledge, but even your beliefs, your look at life and even your values. TED is pure gym for the brain and it was instrumental for drafting the most philosophical passages in this book. I just added a lot of rage, because what we should appreciate about TED is those radical ideas are brought on stage without anger. The speakers are willing to share their ideas, not to attack anyone. Instead, too often we see prophets who prefer to attack random parties instead of explaining in detail why they "hold the truth". I've mentioned TED multiple times over this book, so I take for granted that now you have already subscribed to it.

TED-Ed

An amazing complementary tool to TEDx Talks is TED-Ed. This sub-channel brings to you compelling animated videos explaining complex topics in a simple, and direct manner. Videos are usually short, so perfect for 21st century humans with an attention span not exceeding 5 minutes. I can't go to sleep if I don't watch one TED-Ed video first. I learned so many things which I completely ignored. Why can't we cure cancer? What is exactly schizophrenia? Why is 0 the most important number? Does time really exist? Oh, pure honey for the brain! If I have to suggest a video to start from, I'd say *The origin of countless conspiracy theories*, by Patrick JMT.[105] [106]

It's nothing like you expect. This video examines the mathematical side of the issue. Because our brains are pattern hounds, they will see a pattern everywhere. I don't want to spoil much more. I'll just tell you that if you read Moby Dick and you cross the letters vertically, you will find dire prophecies hidden in it, like the death of Princess Diana! Don't you believe it? Watch the video...

With all these educational tools, I'm very optimistic the next generations will be much smarter than us. I already see how Gen Z kids are much more knowledgeable and flexible than millennials. They are even more immune to conspiracy theories. They are more used to fact checking the bullshits circulating on social media. It's not a coincidence that most conspiracy theories are spread by millennials and Gen X. These generations are much more gullible, maybe because they are less curious and less open minded, having been raised in a world which was more static and with less opportunities to confront ideas. You may disagree and yell that Gen Z are stupid kids posting dull stuff on TikTok all day long. That's just a part of Gen Z, but anyway, does TikTok prove something about Gen Z? Even our generation can brag about years of trash TV where we used to see the same stuff or even worse. Come on! We're not saints either. And still better to watch the stupid stuff on TikTok and Instagram rather than insulting each other over fake news on Facebook. Don't you think?

---

[105]https://www.youtube.com/watch?v=88_C-fogY40
[106]https://www.learnamic.com/resource-providers/patrickjmt

I Will Miss President Corona

In a way or another, I will miss President Corona. It was out of any proportion comparing him to the black plague, to the Spanish Flu, to the biblical locust infestations. The President was regarded as a monster... Maybe he was a monster, but what if the monster was just ugly, but not evil after all...? What if we defeated the wrong enemy, as Eisenhower said at the end of World War II about the German people...?

While writing this book I was listening (with a certain frequency) to an Italian song from 1992. It was written by a brilliant author called Samuele Bersani. *Il Mostro* (The Monster) is one of Bersani's first works, a fine work. The instrumentation is melodic and simple, arranged perfectly on a piano to accompany the touching lyrics that describe exclusion, discrimination and lack of empathy and understanding. I invite you to take a heed and I help you with the meaning by presenting the lyrics hereafter translated in English:

Peering out from a distant world, the last hairy and giant monster
The only example left of six-legged monster
How I like seeing him passing by, what would I do to touch him, what would I do...

They say he's capable of killing a man
not to defend himself, just because he is no good
They say they are accomplished scientists, elite and reliable people
But the only clear thing is that the monster is afraid, the monster is afraid

He's looking for a place far from the evil
sure a cave in the woods would be ideal
But the only quiet place is that old yard
the only space left for a big animal

They say "We're live" the scoop is served
"this is the monster's nest, we've followed him!"
They say they are anchormen, elite and reliable people
But the only clear thing is that the monster is afraid, the monster is afraid

You just need to say that the monster is evil
then you wait a minute and an army comes
bombs and rifles here we are, the attack is total
special forces surround the old yard

They say that they are ready to shoot the monster
"We will get him dead or alive on the spot!"

They say they are action soldiers, elite and reliable people
But the only clear thing is that the monster is afraid, the monster is afraid

The monster would like to hibernate and tries to close his eyes
but he knows that hibernation comes only in winter
He reopens the eyes on the world, this world of monsters,
they are only two-legged, but they are much more monsters

He is left with just one thing
calling his world faraway
He does it with all of his breath, but lower and lower
I wish to save him, take him away on a train
leave him after the rain, there down the rainbow....

This monster gave me a reason to keep on living. He saved my life and who knows how many others. I wouldn't be surprised if we discovered that the lives saved by the coronavirus were more than the ones taken away. Unfortunately, nobody will ever bother to find this out. It doesn't matter. I know why I'm still alive and who I have to thank! Thank You, President Corona! Good-bye...

www.ingramcontent.com/pod-product-compliance
Ingram Content Group UK Ltd.
Pitfield, Milton Keynes, MK11 3LW, UK
UKHW021703190726
13853UKWH00001B/403